EAA Series–Textbook

Editors

D. Filipovic (Co-Chair)   M. F
Ch. Hipp   A. P   r)

EAA series is successor of the EAA Lecture Notes and supported by the European Actuarial Academy (EAA GmbH), founded on the 29 August, 2005 in Cologne (Germany) by the Actuarial Associations of Austria, Germany, the Netherlands and Switzerland. EAA offers actuarial education including examination, permanent education for certified actuaries and consulting on actuarial education.

actuarial-academy.com

For further titles published in this series, please go to
http://www.springer.com/series/7879

Michael Koller

Stochastic Models
in Life Insurance

 Springer

Michael Koller
Mathematics Department
ETH Zürich
Zurich
Switzerland

Translation from the German edition: Michael Koller, Stochastische Modelle in der Lebensversicherung, 2nd Ed. © Springer-Verlag Berlin Heidelberg 2010

ISSN 1869-6929   e-ISSN 1869-6937
ISBN 978-3-642-28438-0   e-ISBN 978-3-642-28439-7
DOI 10.1007/978-3-642-28439-7
Springer Heidelberg New York Dordrecht London

Library of Congress Control Number: 2012933836

Mathematics Subject Classification (2010): 60G35, 62J20, 60J10, 60J27, 60J65, 60K30, 60J70

Printed on acid-free paper

Springer is part of Springer Science+Business Media (www.springer.com)

To Luisa, Giulia and Anna

Preface

This book was originally written in German and goes back to a lecture I held at the ETH in Zurich in 1995. At the time, just after having completed my PhD in abstract mathematics, I was of the opinion that life insurance mathematics would benefit greatly from a mathematically rigid approach, and I decided to write this book in German. Since I very much like to put theory in practice, I also programmed the corresponding algorithms, which are now used at several companies for the calculation of mathematical reserves. This English version of the book is based on the second edition of the original German version and has been expanded: Over the last 15 years since writing the book for the first time, the world has changed considerably, mainly with regard to risk management in insurance companies and also with regard to Solvency II and the computer power now available for models and simulations. In order to reflect these two topics, I have written two additional chapters covering ALM and abstract valuation concepts.

The aim of this book is twofold: On one hand it aims to provide a sound mathematical basis for life insurance mathematics, and on the other it seeks to apply the underlying concepts concretely. This book is mainly written for the advanced mathematical student or actuary and the reader should have a good grasp of basic analysis including measure and integration theory. Moreover the reader should understand basic probability theory and functional analysis. The book is written in a typical mathematical style, in which each theorem is followed by its proof. From a very high level point of view the aim of the book is to provide a robust framework for modelling life insurance policies by means of Markov chains. We will see how to calculate the expected present values and higher moments of future cash flows using recursion formulae. The underlying probability density functions are calculated using a similar approach. In the following chapters unit linked policies with guarantees and insurance portfolios are covered. In the last two chapters of the book we look at abstract valuation concepts and asset liability management.

At this point I would like to thank a few important people, mainly those who helped me to make this book happen and helped me to make it better. My particular thanks go to Hans Bühlmann and Josef Kupper, who motivated me to write this book in the first place in 1995 and also helped me to improve its quality considerably. For the English translation I would like to express my thanks to Bjöern Böttcher. Finally I would also like to thank my colleagues from the different insurance companies (Swiss Life, Swiss Re, Partner Re, Aviva and Prudential) who helped me to better understand insurance and to improve the quality of the book. Finally I would like to thank my wife Luisa and my two daughters Giulia and Anna for their unfailing support.

Herrliberg, 1 September 2011   Michael Koller

Contents

Chapter 1
A General Life Insurance Model

1.1 Introduction

The life insurance market offers a wide range of different policies. It is, without expert knowledge, hardly possible to differentiate between all these policies. This is in particular due to the fact that the content of a life insurance is an abstract good.

A life insurance can always be understood as a bet: either one gets a benefit or one pays the premium without getting anything in return. From this point of view life insurance mathematics is a part of probability theory.

Since a life insurance deals with monetary benefits and premiums it is also part of the financial market and the economy. In this context one should note that insurances whose benefits are unit-linked, e.g. the payout depends on the performance of a fond, actually rely on modern theory of financial markets.

Form a legal point of view a life insurance is a contract between the policy holder and the insurer.

As we have note above, life insurances is characterised by its abstract matter and its diversity. Since its content is abstract its value is not intuitively obvious. This is particular due to the fact that a life insurance is usually only bought once or twice during life time. In contrast, for example one buys a loaf of bread on a regular basis, and thus one has acquired a feeling for its correct value.

A life insurance—in particular an individual policy—is a long term contract. Take for example a 30 year old man who buys a permanent life insurance. Now suppose he dies when he reaches ninety, then the contract period was 60 years.

Due to this long duration of the contract and the risks taken—think for example of changing fundamentals—it is necessary to calculate the price of an insurance with care and foresight.

In this chapter we are going to explain classical types of insurance policies. Furthermore we introduce a general model for life insurance, which can be used to price many of the available policies.

1.2 Examples

We start with a description of the most common types of life insurance and the different methods of financing a policy.

1.2.1 Types of Life Insurance

It is characteristic for every life insurance that the insured event is strongly related to the health of the insured. Thus one can classify life insurance as follows:

- insurance on life or death,
- insurance on permanent disability,
- health insurance.

For an insurance on life or death the essential event is the survival of the insured person up to a certain date or the death before a certain date, respectively. Furthermore these insurance types can be classified by the causes of death which yield a payout (e.g. a life insurance which pays only in the event of an accidental death). Especially, various kinds of survivor's pensions and pure endowments are insurances on life or death.

For a permanent disability insurance the essential criterion is the (dis)ability of the insured at a given date. These insurances have the special feature, that already a certain degree of disability might be sufficient for a claim.

For insurances on health the payout depends on the health of the insured. This class of insurances contains also modern types of policies, like a long term care insurance. The latter only provides benefits if the insured is unable to meet his basic needs (e.g. he is unable to dress himself).

Besides a classification based on the insured event one can also classify the insurance based on the benefit. This can either be paid in annuities or in a lump sum.

In the following we give some typical examples of life insurances.

Pension: A pension policy constitutes that the insurer has to pay annuities to the insured when he reaches a certain age (age of maturity of the policy). Then the pension is paid until the death of the insured. The payment of the annuities is usually done at regular intervals: monthly, quarterly or yearly. Moreover the payment can be done in advance (at the beginning of each interval) or arrears (at the and of each interval). Since the pension is only paid until death, one can additionally agree upon a minimum payment period. In this case the pension is paid at least for the minimum period. (This type of pension contract supplies the desire of the insured to get at least something back for the premiums paid in.)

Pure endowment: A pure endowment insurance provides a payment from the insurer to the insured, if he reaches the age of maturity of the policy. Otherwise there is no payment.

Term/permanent life insurance: A (term/permanent) life insurance is the counterpart to a pure endowment insurance. In contrast to the latter a life insurance does not yield a payout to the insured if the age of maturity of the policy is reached. In the popular case of a term life insurance there is no payout at all if the insured reaches the age of maturity of the policy. But if the insured dies before that age his heirs get a payment. A special case is the permanent life insurance, which yields a payout to the heirs no matter how old the insured is at the time of his death. This insurance is in some countries very popular, since it in a sense an investment into ones off springs.

Endowment: The endowment insurance is the classic example of a life insurance. It is the sum of a pure endowment insurance and a term or permanent life insurance. This means that it yields a payout in the case of an early death and also in the case of reaching the fixed age of maturity.

Widow's pension: A widow's pension is connected to the life of two persons. This is in contrast to the previous examples, where only the life of one person was considered. For a widow's pension there is the insured (the person whose life is insured, e.g. husband) and the beneficiary person (e.g. spouse). As long as both persons are alive no payment is due. If the insured dies and the beneficiary is still alive, then the beneficiary gets a pension until death. Also for this kind of insurance it is possible to fix a minimum period of payments in the policy.

Orphan's pension: After the death of the father or the mother their child gets a pension until it is of age or until death.

Insurance on two lives: For an insurance on two lives, as in the case of a widow's or orphan's pension, one has to consider the lives of two persons. Here the policy fixes a payment depending on the state of the two persons (insured, beneficiary)$\in$ $\{(**), (*†), (†*), (††)\}$. Obviously, a widow's or an orphan's pension is just a special case of the insurance on two lives. As before, also in this case one could agree upon a minimum period of payments.

Refund guarantee: The refund guarantee is an additional insurance which is often sold together with a pension or a pure endowment. It is a life insurance whose payment equals the paid-in premiums, possibly reduced by the already received payments. The refund guarantee supplies the same want as the minimum periods of payment for the pensions.

Now we have discussed the main insurance types on life and death. Next we want to give a short description of insurances on permanent disability. For these the ability to work is the main criterion. In this context one should note that the probability of becoming disabled strongly depends on the economic environment. This is due to the fact that in a good economic environment everyone finds a job. But a person with restricted health has a hard time finding a job during an economic downturn. In connection with disability the following types of insurances are most common:

Disability pension: In the case of disability, after an initial waiting period, the insured gets a pension until he reaches a fixed age (or until his death) or until he is able to work again. A disability pension without fixed age for a final payment is called permanent disability pension. Often an initial waiting period is introduced, since in most cases disability occurs after an accident or illness and the person actually recovers quickly thereafter. Thus the initial waiting period reduces the price of these policies. Typical waiting periods are three or six month and one or two years.

Disability capital: The disability capital insurance pays a lump sum to the insured in case of a permanent disability.

Premium waiver: The premium waiver is an additional insurance. It waives the obligation to pay further premiums for the insured in the case of disability. Waiting periods are also common for this type of insurance.

Disability children's pension: This pension is similar to the orphan's pension. The only difference is that the cause for a payment is the disability of the mother or father instead of their death.

1.2.2 Methods of Financing

Previously we looked at the different types of insurances. Now we are going to discuss the different financing methods and the ideas they are based on. The main principle of life insurances states that the value of the benefits provided by the insurer is equivalent to the value of the policy for the insured. Obviously one has to discuss this equivalence relation in more detail. We will do this in the next chapters and provide a precise definition of the equivalence principle. Then it will be used to calculate the premiums, but for now we get back to the methods of financing. The two common types are:

- Financing by premiums,
- Financing by a single payment (single premium).

Financing by premiums requires the insured to pay premiums to the insurer at regular intervals. This obligation usually ends either when the age of maturity of the policy is reached or if the insured dies.

The other option to finance a life insurance is a single payment. Often a policy incorporates a mixture of both financing methods.

1.3 The Insurance Model

In this section we want to introduce the insurance model which we will use thereafter. We attempt to describe the real world by a model. Thus it is important to use a model class which is flexible enough to accommodate this. Figure 1.1 shows the general setup of an insurance model. Here we think of an insured person who, at every time t, is in a state $1, 2, \ldots, n$. State 1 could for example indicate that the

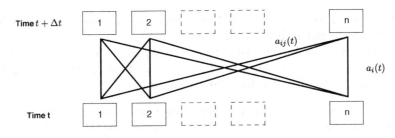

Fig. 1.1 Policy setup from t to $t + \Delta t$

person is alive. The state of the person is then given by the stochastic process X with $X_t(\omega) \in S = \{1, 2, \ldots, n\}$. When the insured remains in one state or switches its state a payment, as defined in the insurance policy, is due. For this there are functions $a_i(t)$ and $a_{ij}(t)$ given which correspond to the lines in the figure. They define the amount which the insured gets if he remains in state i (payment $a_i(t)$) or if he switches from state i to j at time t (payment $a_{ij}(t)$). In the following we are going to introduce the necessary concepts for this setup. One distinguishes between the continuous time model, where $(X_t)_{t \in T}$ is defined on an interval in $\mathbb{R}$, and the discrete time model, where $(X_t)_{t \in T}$ is defined on a subset of $\mathbb{N}$. The continuous time model yields the more interesting statements whereas the discrete model is very important in applications, therefore we will discuss both models.

Definition 1.3.1 (*State space*) We denote by S the state space which is used for the insurance policy. S is finite set.

Example 1.3.2 For a life insurance or an endowment one often uses the state space $S = \{*, \dagger\}$.

Example 1.3.3 For a disability insurance one has to consider at least the states: alive (active), dead and disabled. Often one uses more states to get a better model. For example in Switzerland a model is used which uses the states $\{*, \dagger\}$ and the family of states $\{$person became disabled at the age of $x : x \in \mathbb{N}\}$.

With the states defined, it is now possible to derive a mathematical model for the payments/benefits. To define the model of the benefits, the so called policy functions, it is necessary to define the time set more precisely. We will define two different time sets since a discrete time set is often used in applications, but the continuous time set yields the neater results.

Definition 1.3.4 • $a_i(t)$ denotes the sum of the payments to the insured up to time t, given that we know that he has always been in state i. The $a_i(t)$ are called the generalised pension payments. If this pension function is of bounded variation (see Definition 2.1.5) we can also write $a_i(t) = \int_0^t da_i(s)$.
• $a_{ij}(t)$ denotes the payments which are due when the state switches from i to j at time t. These benefits are called generalised capital benefits.

- In the case of a discrete time set $a_i^{\text{Pre}}(t)$ denotes the pension payment which is due at time t, given that the insured is at time t in i.
- In the case of a discrete time set $a_{ij}^{\text{Post}}(t)$ denotes the capital benefits which are due when switching from i at time t to j at time $t+1$. We are going to assume that the payment is transferred at the end of the time interval.

The functions $a_i(t)$ are different in the continuous time model and the discrete time model. In the former $a_i(t)$ denotes the sum of the pension payments which are payed up to time t, similar to a mileage meter in a car. In the latter $a_i^{\text{Pre}}(t)$ denotes the single pension payment at time t.

The following example illustrates the interplay between the state space and the functions which define the policy.

Example 1.3.5 Consider an endowment policy with 200,000 USD death benefit and 100,000 USD survival benefit. This insurance shall be financed by a yearly premium of 2,000 USD.

For a age at maturity of 65 the non trivial policy functions are:

$$
a_*(x) = \begin{cases} 0, & \text{if } x < x_0, \\ -\int_{x_0}^x 2000\, dt, & \text{if } x \in [x_0, 65], \\ -(65 - x_0) \times 2000 + 100000, & \text{if } x > 65, \end{cases}
$$

$$
a_{*\dagger}(x) = \begin{cases} 0, & \text{if } x < x_0 \text{ or } x > 65, \\ 200000, & \text{if } x \in [x_0, 65], \end{cases}
$$

where x_0 is the age of entry into the contract, $*$ and $\dagger$ denote the states alive and dead, respectively.

Chapter 2
Stochastic Processes

2.1 Definitions

In this section we will recall basic definitions from probability theory. These will be used throughout the book.

To understand this chapter a basic knowledge in probability theory, measure theory and analysis is a prerequisite.

Definition 2.1.1 (*Sets*) We are going to use the notations:

$$\mathbb{N} = \text{the set of the natural numbers including } 0,$$
$$\mathbb{N}_+ = \{x \in \mathbb{N} : x > 0\},$$
$$\mathbb{R} = \text{the set of the real numbers},$$
$$\mathbb{R}_+ = \{x \in \mathbb{R} : x \geq 0\}.$$

Furthermore we use the following notations for intervals. For $a, b \in \mathbb{R}, a < b$ we write

$$[a, b] := \{t \in \mathbb{R} : a \leq t \leq b\},$$
$$]a, b] := \{t \in \mathbb{R} : a < t \leq b\},$$
$$]a, b[:= \{t \in \mathbb{R} : a < t < b\},$$
$$[a, b[:= \{t \in \mathbb{R} : a \leq t < b\}.$$

Definition 2.1.2 (*Indicator function*) For $A \subset \Omega$ we define the indicator function $\chi_A : \Omega \to \mathbb{R}, \omega \mapsto \chi_A(\omega)$ by

$$\chi_A(\omega) := \begin{cases} 1, & \text{if } \omega \in A, \\ 0, & \text{if } \omega \notin A. \end{cases}$$

Furthermore δ_{ij} is Kronecker's delta, i.e., it is equal to 1 for $i = j$ and 0 otherwise.

M. Koller, *Stochastic Models in Life Insurance*, EAA Series,
DOI: 10.1007/978-3-642-28439-7_2, © Springer-Verlag Berlin Heidelberg 2012

Definition 2.1.3 Let

$$f : \mathbb{R} \to \mathbb{R}, x \mapsto f(x).$$

We define, if they exist, the left limit and the right limit of f at x by:

$$f(x^-) := \lim_{\xi \uparrow x} f(\xi),$$

$$f(x^+) := \lim_{\xi \downarrow x} f(\xi).$$

Definition 2.1.4 A real valued function $f : \mathbb{R} \to \mathbb{R}$ is said to be of order $o(t)$, if

$$\lim_{t \to 0} \frac{f(t)}{t} = 0.$$

This is denoted by $f(t) = o(t)$.

Definition 2.1.5 (*Function of bounded variation*) Let $I \subset \mathbb{R}$ be a bounded interval. For a function

$$f : I \to \mathbb{R}, t \mapsto f(t)$$

the total variation of the function f on the interval I is defined by

$$V(f, I) = \sup \sum_{i=1}^{n} |f(b_i) - f(a_i)|,$$

where the supremum is taken with respect to all partitions of the interval I satisfying

$$a_1 \leq b_1 \leq a_2 \leq b_2 \leq \cdots \leq a_n \leq b_n.$$

The function f is of *bounded variation* on I, if $V(f, I)$ is finite. Functions corresponding to a life insurance are usually defined on the interval $[0, \omega]$, where $\omega < \infty$ denotes the last age at which some individuals are alive.

Properties of functions of bounded variation can be found for example in [DS57].

It is important to note, that functions of bounded variation form an algebra and a lattice. Thus, if f, g are functions of bounded variation and $\alpha \in \mathbb{R}$, then the following functions are also of bounded variation: $\alpha f + g$, $f \times g$, $\min(0, f)$ and $\max(0, f)$.

Definition 2.1.6 (*Probability space, stochastic process*) We denote by $(\Omega, \mathcal{A}, P)$ a probability space which satisfies Kolmogorov's axioms.

Let $(S, \mathcal{S})$ be a measurable space (i.e. S is a set and $\mathcal{S}$ is a σ-algebra on S) and T be a set.

The Borel σ-algebra on the real numbers will be denoted by $\mathcal{R} = \sigma(\mathbb{R})$.

A family $\{X_t : t \in T\}$ of random variables

$$X_t : (\Omega, \mathcal{A}, P) \to (S, \mathcal{S}), \omega \mapsto X_t(\omega)$$

is called *stochastic process* on $(\Omega, \mathcal{A}, P)$ with *state space S*.

For each $\omega \in \Omega$ a sample path of the process is given by the function

$$X.(\omega) : T \to S, t \mapsto X_t(\omega).$$

We assume that each sample path is right continuous and has left limits.

Definition 2.1.7 (*Expectations*) Let X be a random variable on $(\Omega, \mathcal{A}, P)$ and $\mathcal{B} \subset \mathcal{A}$ be a σ-algebra. Then we denote by

- $E[X]$ the *expectation* of the random variable X,
- $V[X]$ the *variance* of the random variable X,
- $E[X|\mathcal{B}]$ the *conditional expectation* of X with respect to $\mathcal{B}$.

Definition 2.1.8 Let $(X_t)_{t\in T}$ be a stochastic process on $(\Omega, \mathcal{A}, P)$ taking values in a countable set S. We define for $j \in S$ the indicator function with respect to the process $(X_t)_{t\in T}$ at time t by

$$I_j(t)(\omega) = \begin{cases} 1, & \text{if } X_t(\omega) = j, \\ 0, & \text{if } X_t(\omega) \neq j. \end{cases}$$

Analogous, we define for $j, k \in S$ the number of jumps from j to k in the time interval $]0, t[$ by

$$N_{jk}(t)(\omega) = \#\{\tau \in]0, t[: X_{\tau^-} = j \text{ and } X_\tau = k\}.$$

Remark 2.1.9 In the following the function $I_j(t)$ is used to check if the insured person is at time t in state j. Thus one can check if the pension $a_j(t)$ has to be paid. Similarly, a switch from i to j is indicated by an increase of $N_{ij}(t)$ by 1.

Definition 2.1.10 (*Normal distribution*) A random variable X on $(\mathbb{R}, \sigma(\mathbb{R}))$ with density

$$f_{\mu,\sigma^2}(x) = \frac{1}{\sqrt{2\pi\sigma^2}} \exp\left(-\frac{(x-\mu)^2}{2\sigma^2}\right), \qquad x \in \mathbb{R}$$

is called *normal distributed* with expectation μ and variance σ^2. Such a random variable is denoted by $X \sim \mathcal{N}(\mu, \sigma^2)$.

Examples of stochastic processes are:

Example 2.1.11 (Brownian motion) An example of a non trivial stochastic process is Brownian motion. Brownian motion $X = (X_t)_{t\geq 0}$ with continuous time set $(T = \mathbb{R}_+)$ and state space $S = \mathbb{R}$ is used to model many real world phenomena.

The process is characterised by the following properties:

1. $X_0 = 0$ almost surely.
2. X has independent increments: the random variables $B_{t_1} - B_{t_0}$, $B_{t_2} - B_{t_1}$, ..., $B_{t_n} - B_{t_{n-1}}$ are independent for all $0 \leq t_1 < t_2 < \cdots < t_n$ and all $n \in \mathbb{N}$.
3. X has stationary increments.
4. $X_t \sim \mathcal{N}(0, t)$.

One can show that almost all sample paths of X are continuous and nowhere differentiable.

Example 2.1.12 (Poisson process) The Poisson process $N = (N_t)_{t \geq 0}$ is a *counting process* with state space $\mathbb{N}$. For example it is used in insurance mathematics to model the number of incurred claims. This process also uses a continuous time set. The homogeneous Poisson process is characterised by the following properties:

1. $N_0 = 0$ almost surely.
2. N has independent and
3. N stationary increments.
4. For all $t > 0$ and all $k \in \mathbb{N}$ gilt: $P[N_t = k] = \exp(-\lambda t) \frac{(\lambda t)^k}{k!}$.

2.2 Markov Chains on a Countable State Space

In the following S is a countable set.

Definition 2.2.1 Let $(X_t)_{t \in T}$ be a stochastic process on $(\Omega, \mathcal{A}, P)$ with state space S and $T \subset \mathbb{R}$. The process X is called Markov chain, if for all

$$n \geq 1, \ t_1 < t_2 < \cdots < t_{n+1} \in T, \ i_1, i_2, \ldots, i_{n+1} \in S$$

with

$$P[X_{t_1} = i_1, X_{t_2} = i_2, \ldots, X_{t_n} = i_n] > 0$$

the following statement holds:

$$P[X_{t_{n+1}} = i_{n+1} | X_{t_k} = i_k \forall k \leq n] = P[X_{t_{n+1}} = i_{n+1} | X_{t_n} = i_n]. \qquad (2.1)$$

Remark 2.2.2 1. Equation (2.1) states that the conditional probabilities only depend on the *last* state. They do not depend on the path which led the chain into that state.
2. Markov chains are very versatile in their applications. This is due to the fact, that on the one hand they are very easy to handle and on the other hand they can model a wide range of phenomena. In the following we are going to model life insurances by Markov chains.

Example 2.2.3   1. Let $(X_t)_{t \in T}$ be a stochastic process with $S \subset \mathbb{R}$ and $T = \mathbb{N}_+$, for which the random variables $\{X_t : t \in T\}$ are independent. This process is a Markov chain since

$$P[X_{t_1} = i_1, X_{t_2} = i_2, \ldots, X_{t_n} = i_n] = \prod_{k=1}^{n} P[X_{t_k} = i_k]$$

for $n \geq 1$, $t_1 < t_2 < \cdots < t_{n+1} \in T$, $i_1, i_2, \ldots, i_{n+1} \in S$.

2. Based on the previous example we define $S_m = \sum_{k=1}^{m} X_k$, where $m \in \mathbb{N}$. This is also an example of a Markov chain.

Proof

$$\begin{aligned} P[S_{t_{n+1}} = i_{n+1} \mid S_{t_1} = i_1, S_{t_2} = i_2, \ldots, S_{t_n} = i_n] \\ = P[S_{t_{n+1}} - S_{t_n} = i_{n+1} - i_n] \\ = P[S_{t_{n+1}} = i_{n+1} \mid S_{t_n} = i_n]. \end{aligned}$$

Definition 2.2.4 Let $(X_t)_{t \in T}$ be a stochastic process on $(\Omega, \mathcal{A}, P)$. Then

$$p_{ij}(s, t) := P[X_t = j \mid X_s = i], \quad \text{where } s \leq t \text{ and } i, j \in S,$$

is called the conditional probability to switch from state i at time s to state j at time t, or also transition probability for short.

The following theorem of Chapman and Kolmogorov is fundamental for the theory which we will present in the next chapters. The theorem states the relation of $P(s, t)$, $P(t, u)$ and $P(s, u)$ for $s \leq t \leq u$.

Theorem 2.2.5 (Chapman–Kolmogorov equation) *Let $(X_t)_{t \in T}$ be a Markov chain. For $s \leq t \leq u \in T$ and $i, k \in S$ such that $P[X_s = i] > 0$ the following equations hold:*

$$p_{ik}(s, u) = \sum_{j \in S} p_{ij}(s, t)\, p_{jk}(t, u), \tag{2.2}$$

$$P(s, u) = P(s, t) \times P(t, u). \tag{2.3}$$

This shows, that one can get $P(s, u)$ by matrix multiplication of $P(s, t)$ and $P(t, u)$ for $s \leq t \leq u \in T$.

Proof Obviously, the equation holds for $t = s$ or $t = u$. Thus we can assume $s < t < u$ without loss of generality. We will use the following notation:

$$S^* = \{j \in S : P[X_t = j \mid X_s = i] \neq 0\}$$
$$= \{j \in S : P[X_t = j, X_s = i] \neq 0\}.$$

(The last equality holds since $P[X_s = i] > 0$.) Now the Chapman–Kolmogorov equation can be deduced from the following equation:

$$
\begin{aligned}
p_{ik}(s, u) &= P[X_u = k \mid X_s = i] \\
&= \sum_{j \in S^*} P[X_u = k, X_t = j \mid X_s = i] \\
&= \sum_{j \in S^*} P[X_t = j \mid X_s = i] \times P[X_u = k \mid X_s = i, X_t = j] \\
&= \sum_{j \in S^*} p_{ij}(s, t) \times p_{jk}(t, u) \\
&= \sum_{j \in S} p_{ij}(s, t) \times p_{jk}(t, u),
\end{aligned}
$$

where we applied the Markov property to get equality in the forth line.

After proving the Chapman–Kolmogorov equation we are now able to introduce the abstract concept of transition matrices.

Definition 2.2.6 (*Transition matrix*) A family $(p_{ij}(s, t))_{(i,j) \in S \times S}$ is called transition matrix, if the following four properties hold:

1. $p_{ij}(s, t) \geq 0$.
2. $\sum_{j \in S} p_{ij}(s, t) = 1$.
3. $p_{ij}(s, s) = \begin{cases} 1, & \text{if } i = j, \\ 0, & \text{if } i \neq j, \end{cases} \quad$ if $P[X_s = i] > 0$.
4. $p_{ik}(s, u) = \sum_{j \in S} p_{ij}(s, t) p_{jk}(t, u)$ for $s \leq t \leq u$ and $P[X_s = i] > 0$.

Theorem 2.2.7 *Let* $(X_t)_{t \in T}$ *be a Markov chain. Then* $(p_{ij}(s, t))_{(i,j) \in S \times S}$ *is a transition matrix.*

Proof This theorem is a direct consequence of the theorem by Chapman and Kolmogorov (Theorem 2.2.5).

Theorem 2.2.8 *A stochastic process* $(X_t)_{t \in T}$ *is a Markov chain, if and only if*

$$P[X_{t_1} = i_1, \ldots, X_{t_n} = i_n] = P[X_{t_1} = i_1] \prod_{k=1}^{n-1} p_{i_k, i_{k+1}}(t_k, t_{k+1}), \qquad (2.4)$$

for all

$$n \geq 1, \ t_1 < t_2 < \cdots < t_{n+1} \in T, \ i_1, i_2, \ldots, i_{n+1} \in S.$$

Proof Let $(X_t)_{t \in T}$ be a Markov chain satisfying

$$P[X_{t_1} = i_1, X_{t_2} = i_2, \dots, X_{t_n} = i_n] > 0.$$

Then the Markov property implies

$$P[X_{t_1} = i_1, \dots, X_{t_n} = i_n] = P[X_{t_1} = i_1, \dots, X_{t_{n-1}} = i_{n-1}] \cdot p_{i_{n-1}, i_n}(t_{n-1}, t_n).$$

This yields (2.4) by induction. The converse statement is trivial.

Theorem 2.2.9 (Markov property) *Let* $(X_t)_{t \in T}$ *be a Markov chain and* n, m *be elements of* $\mathbb{N}$. *Fix* $t_1 < t_2 < \cdots < t_n < t_{n+1} < \cdots < t_{n+m}$, $i \in S$ *and sets* $A \subset S^{n-1}$ *(where* S^{n-1} *denotes the* $n-1$ *times Cartesian product of the set* S*) and* $B \subset S^m$ *such that*

$$P\left[(X_{t_1}, X_{t_2}, \dots, X_{t_{n-1}}) \in A, X_{t_n} = i\right] > 0.$$

Then the following equation (Markov property) *holds:*

$$P\left[(X_{t_{n+1}}, X_{t_{n+2}}, \dots, X_{t_{n+m}}) \in B \mid (X_{t_1}, X_{t_2}, \dots, X_{t_{n-1}}) \in A, X_{t_n} = i\right]$$
$$= P\left[(X_{t_{n+1}}, X_{t_{n+2}}, \dots, X_{t_{n+m}}) \in B \mid X_{t_n} = i\right].$$

Proof We use the notation $i^n = (i_1, i_2, \dots, i_n)$. An application of Eq. (2.4) yields:

$$P\left[(X_{t_1}, \dots, X_{t_{n-1}}) \in A, X_{t_n} = i\right]$$
$$= \sum_{i^{n-1} \in A, i_n = i} P\left[X_{t_1} = i_1\right] \times \prod_{k=1}^{n-1} p_{i_k, i_{k+1}}(t_k, t_{k+1}),$$
$$P\left[(X_{t_1}, \dots, X_{t_{n+m}}) \in A \times \{i\} \times B\right]$$
$$= \sum_{i^{n+m} \in A \times \{i\} \times B} P\left[X_{t_1} = i_1\right] \times \prod_{k=1}^{n+m-1} p_{i_k, i_{k+1}}(t_k, t_{k+1}).$$

Finally these two equations imply

$$P\left[(X_{t_{n+1}}, \dots, X_{t_{n+m}}) \in B \mid (X_{t_1}, \dots, X_{t_{n-1}}) \in A, X_{t_n} = i\right]$$
$$= \sum_{(i_n, i_{n+1}, \dots, i_{n+m}) \in \{i\} \times B} \prod_{k=n}^{n+m-1} p_{i_k, i_{k+1}}(t_k, t_{k+1})$$
$$\times \frac{\sum_{i^{n-1} \in A} P\left[X_{t_1} = i_1\right] \times \prod_{l=1}^{n-1} p_{i_l, i_{l+1}}(t_l, t_{l+1})}{\sum_{i^{n-1} \in A} P\left[X_{t_1} = i_1\right] \times \prod_{l=1}^{n-1} p_{i_l, i_{l+1}}(t_l, t_{l+1})}$$

$$= \sum_{(i_{n+1},\ldots,i_{n+m})\in B} \prod_{k=n}^{n+m-1} p_{i_k,i_{k+1}}(t_k,t_{k+1}) \frac{P\left[X_{t_n}=i\right]}{P\left[X_{t_n}=i\right]}$$

$$= P\left[(X_{t_{n+1}}, X_{t_{n+2}}, \ldots, X_{t_{n+m}}) \in B | X_{t_n} = i\right].$$

Definition 2.2.10 A Markov chain $(X_t)_{t\in T}$ is called *homogeneous*, if it is time homogeneous, i.e., the following equation holds for all $s, t \in \mathbb{R}, h > 0$ and $i, j \in S$ such that $P[X_s = i] > 0$ and $P[X_t = i] > 0$:

$$P[X_{s+h} = j \mid X_s = i] = P[X_{t+h} = j \mid X_t = i].$$

For a homogeneous Markov chain we use the notation:

$$p_{ij}(h) := p_{ij}(s, s+h),$$
$$P(h) := P(s, s+h).$$

Remark 2.2.11   1. A homogeneous Markov chain is characterised by the fact, that the transition probabilities, and therefore also the transition matrices, only depend on the size of the time increment.
2. For a homogeneous Markov chain one can simplify the Chapman–Kolmogorov equations to the semi-group property:

$$P(s + t) = P(s) \times P(t).$$

The semi-group property is popular in many different areas e.g. in quantum mechanics.
3. The mapping
$$P : T \to M_n(\mathbb{R}), t \mapsto P(t)$$

defines a one parameter *semi-group*.

2.3 Markov Chains in Continuous Time and Kolmogorov's Differential Equations

In the following we will only consider Markov chains on a finite state space. Thus point wise convergence and uniform convergence will coincide on S. This enables us to give some of the proofs in a simpler form.

Definition 2.3.1 Let $(X_t)_{t\in T}$ be a Markov chain with finite state space S and $T \subset \mathbb{R}$. For $N \subset S$ we define
$$p_{jN}(s, t) := \sum_{k\in N} p_{jk}(s, t).$$

Definition 2.3.2 (*Transition rates*) Let $(X_t)_{t \in T}$ be a Markov chain in continuous time with finite state space S. $(X_t)_{t \in T}$ is called *regular*, if

$$\mu_i(t) = \lim_{\Delta t \searrow 0} \frac{1 - p_{ii}(t, t + \Delta t)}{\Delta t} \quad \text{for all } i \in S, \tag{2.5}$$

$$\mu_{ij}(t) = \lim_{\Delta t \searrow 0} \frac{p_{ij}(t, t + \Delta t)}{\Delta t} \quad \text{for all } i \neq j \in S \tag{2.6}$$

are well defined and continuous with respect to t.

The functions $\mu_i(t)$ and $\mu_{ij}(t)$ are called *transition rates* of the Markov chain. Furthermore we define μ_{ii} by

$$\mu_{ii}(t) = -\mu_i(t) \text{ for all } i \in S. \tag{2.7}$$

Remark 2.3.3 1. In the insurance model the regularity of the Markov chain is used to derive the differential equations which are satisfied by the mathematical reserve corresponding to the policy (Thiele's differential equation, e.g. Theorem 5.2.1).
2. One can understand the transition rates as derivatives of the transition probabilities. For example we get for $i \neq j$:

$$\begin{aligned} \mu_{ij}(t) &= \lim_{\Delta t \searrow 0} \frac{p_{ij}(t, t + \Delta t)}{\Delta t} \\ &= \lim_{\Delta t \searrow 0} \frac{p_{ij}(t, t + \Delta t) - p_{ij}(t, t)}{\Delta t} \\ &= \frac{d}{ds} p_{ij}(t, s) \Big|_{s=t}. \end{aligned}$$

3. $\mu_{ij}(t)\,dt$ can be understood as the infinitesimal transition rate from i to j ($i \rightsquigarrow j$) in the time interval $[t, t + dt]$. Similarly, $\mu_i(t)\,dt$ can be understood as the infinitesimal probability of leaving state i in the corresponding time interval. Let us define

$$\Lambda(t) = \begin{pmatrix} \mu_{11}(t) & \mu_{12}(t) & \mu_{13}(t) & \cdots & \mu_{1n}(t) \\ \mu_{21}(t) & \mu_{22}(t) & \mu_{23}(t) & \cdots & \mu_{2n}(t) \\ \mu_{31}(t) & \mu_{32}(t) & \mu_{33}(t) & \cdots & \mu_{3n}(t) \\ \vdots & \vdots & \vdots & \ddots & \vdots \\ \mu_{n1}(t) & \mu_{n2}(t) & \mu_{n3}(t) & \cdots & \mu_{nn}(t) \end{pmatrix}.$$

In a sense, Λ generates the behaviour of the Markov chain. That is, for a homogeneous Markov chain the following equation holds:

$$\Lambda(0) = \lim_{\Delta t \to 0} \frac{P(\Delta t) - 1}{\Delta t}.$$

$\Lambda := \Lambda(0)$ is called the *generator of the one parameter semi group*. We can reconstruct $P(t)$ by

$$P(t) = \exp(t\,\Lambda) = \sum_{n=0}^{\infty} \frac{t^n}{n!} \Lambda^n.$$

4. In the remainder of the book we will only consider finite state spaces. This enables us to avoid certain technical difficulties with respect to convergence.

Based on the transition rates we can prove Kolmogorov's differential equations. These connect the partial derivatives of p_{ij} with μ:

Theorem 2.3.4 (Kolmogorov) *Let $(X_t)_{t \in T}$ be a regular Markov chain on a finite state space S. Then the following statements hold:*

1. *(Backward differential equations)*

$$\frac{d}{ds} p_{ij}(s, t) = \mu_i(s) p_{ij}(s, t) - \sum_{k \neq i} \mu_{ik}(s) p_{kj}(s, t), \qquad (2.8)$$

$$\frac{d}{ds} P(s, t) = -\Lambda(s) P(s, t). \qquad (2.9)$$

2. *(Forward differential equations)*

$$\frac{d}{dt} p_{ij}(s, t) = -p_{ij}(s, t) \mu_j(t) + \sum_{k \neq j} p_{ik}(s, t) \mu_{kj}(t), \qquad (2.10)$$

$$\frac{d}{dt} P(s, t) = P(s, t) \Lambda(t). \qquad (2.11)$$

Proof The major part of the proof is based on the equations of Chapman and Kolmogorov.

1. We will prove the matrix version of the statement. This will help to highlight the key properties. Let $\Delta s > 0$ and set $\xi := s + \Delta s$.

$$\begin{aligned}
\frac{P(\xi, t) - P(s, t)}{\Delta s} &= \frac{1}{\Delta s}\left(P(\xi, t) - P(s, \xi)\,P(\xi, t) \right) \\
&= \left(\frac{1}{\Delta s}(1 - P(s, \xi)) \right) \times P(\xi, t) \\
&\longrightarrow -\Lambda(s) P(s, t) \text{ for } \Delta s \searrow 0,
\end{aligned}$$

where we used the Chapman–Kolmogorov equation and the continuity of the matrix multiplication.

2. Analogous one can prove the forward differential equation. Let $\Delta t > 0$.

$$\frac{P(s, t + \Delta t) - P(s, t)}{\Delta t} = \frac{1}{\Delta t}\left(P(s,t)P(t, t + \Delta t) - P(s,t) \right)$$

$$= P(s,t) \times \frac{1}{\Delta t}\left(P(t, t + \Delta t) - 1 \right)$$

$$\longrightarrow P(s,t)\Lambda(t) \text{ for } \Delta t \searrow 0.$$

Remark 2.3.5 The primary application of Kolmogorov's differential equations is to calculate the transition probabilities p_{ij} based on the rates μ.

Definition 2.3.6 Let $(X_t)_{t \in T}$ be a regular Markov chain on a finite state space S. Then we denote the conditional probability to stay during the time interval $[s, t]$ in j by

$$\bar{p}_{jj}(s, t) := P\left[\bigcap_{\xi \in [s,t]} \{X_\xi = j\} \,\big|\, X_s = j \right]$$

where $s, t \in \mathbb{R}$, $s \le t$ and $j \in S$.

In the setting of a life insurance this probability can for example be used to calculate the probability that the insured survives 5 years. The following theorem illustrates how this probability can be calculated based on the transition rates.

Theorem 2.3.7 *Let $(X_t)_{t \in T}$ be a regular Markov chain. Then*

$$\bar{p}_{jj}(s, t) = \exp\left(-\sum_{k \neq j} \int_s^t \mu_{jk}(\tau)d\tau \right) \tag{2.12}$$

holds for $s \le t$, if $P[X_s = j] > 0$.

Proof We define $K_j(s, t)$ by $K_j(s, t) := \bigcap_{\xi \in [s,t]}\{X_\xi = j\}$. Let $\Delta t > 0$. We have $P[A \cap B \,|\, C] = P[B \,|\, C]\,P[A \,|\, B \cap C]$ and thus

$$\bar{p}_{jj}(s, t + \Delta t) = P[K_j(s, t) \cap K_j(t, t + \Delta t) \,|\, X_s = j]$$
$$= P[K_j(s, t) \,|\, X_s = j]\,P[K_j(t, t + \Delta t) \,|\, X_s = j \cap K_j(s, t)]$$
$$= P[K_j(s, t) \,|\, X_s = j]\,P[K_j(t, t + \Delta t) \,|\, X_t = j]$$
$$= \bar{p}_{jj}(s, t)\,P[K_j(t, t + \Delta t) \,|\, X_t = j],$$

where we used the Markov property and the relation $\{X_s = j\} \cap K_j(s, t) = \{X_t = j\} \cap K_j(s, t)$. The previous equation yields

$$\bar{p}_{jj}(s, t + \Delta t) - \bar{p}_{jj}(s, t) = -\bar{p}_{jj}(s, t) \times \left(1 - P[K_j(t, t + \Delta t) \mid X_t = j]\right)$$

$$= -\bar{p}_{jj}(s, t) \times \left(\sum_{k \neq j} p_{jk}(t, t + \Delta t) + o(\Delta t)\right),$$

where we used that the rates $\mu_{..}$ are well defined.
Now taking the limit we get the differential equation

$$\frac{d}{dt}\bar{p}_{jj}(s, t) = -\bar{p}_{jj}(s, t) \times \sum_{k \neq j} \mu_{jk}(t).$$

Solving this equation with the boundary condition $\bar{p}_{jj}(s, s) = 1$ yields the statement of the theorem, (2.12).

2.4 Examples

In this section we want to illustrate the theory of the previous sections by some examples.

Example 2.4.1 (Life insurance) We start with a life insurance, which provides a sum of money to the heirs in case of the death of the insured. Usually one uses for this a model with either two states (alive *, dead †) or three states (alive, dead (accident), dead (disease)). We will use the model with two states, and the death rate will be exemplary modelled by the function

$$\mu_{*\dagger}(x) = \exp(-9.13275 + 8.09438 \cdot 10^{-2}x - 1.10180 \cdot 10^{-5}x^2). \quad (2.13)$$

The death rate is the transition rate of the state transition $* \rightsquigarrow \dagger$. See Sect. 4.3 for a derivation of the death rate. Based on the death rate and formula (2.12) we are now able to calculate the survival probability of a 35 year old man:

$$\bar{p}_{**}(35, x) = \exp\left(-\int_{35}^{x} \mu_{*\dagger}(\tau)d\tau\right), \qquad \text{for } x > 35.$$

Figure 2.1 shows the transition rate (dotted line) and the survival probability (continuous line) based on $x = 35$.

Example 2.4.2 (Disability pension) We consider a model of a disability pension with the following three states:

State	Symbol
Active	*
Disabled	◇
Dead	†

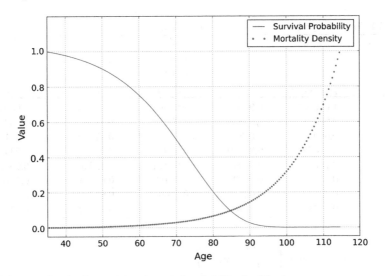

Fig. 2.1 Mortality density $\mu_{*\dagger}(x)$ and survival probability $\bar{p}_{**}(35, x)$

The transition rates are defined by

$$\sigma(x) := 0.0004 + 10^{(0.060\,x - 5.46)},$$
$$\mu(x) := 0.0005 + 10^{(0.038\,x - 4.12)},$$
$$\mu_{*\diamond}(x) := \sigma(x),$$
$$\mu_{*\dagger}(x) := \mu(x),$$
$$\mu_{\diamond\dagger}(x) := \mu(x).$$

The transition rate σ is the infinitesimal probability of becoming disabled and μ is the corresponding probability of dying. We set the other transition rates equal to 0. Thus in particular this model does not incorporate the possibility of becoming active again ($\mu_{\diamond*} = 0$). Moreover one should note that in this model the mortality of disabled persons is equal to the mortality of active persons. This is a simplification, since in reality disabled persons have a higher mortality (they die earlier with a higher probability) than active persons. Therefore, this model yields an overpriced premium for the disability pension.

The explicit knowledge of the transition probabilities p_{ij} is useful for many formulas in insurance mathematics. For the current model they can be calculated by Kolmogorov's differential equations. We get

$$p_{**}(x, y) = \exp\left(-\int_x^y [\mu(\tau) + \sigma(\tau)]\,d\tau\right),$$
$$p_{*\diamond}(x, y) = \exp\left(-\int_x^y \mu(\tau)d\tau\right) \times \left(1 - \exp\left(-\int_x^y \sigma(\tau)d\tau\right)\right),$$

Table 2.1 Transition probabilities for the disability insurance

Initial age			$x_0 = 30$		
Algorithm			Runge-Kutta of order 4		
Step width			0.001		
Age x	$p_{**}(x_0, x)$	$p_{*\diamond}(x_0, x)$	$p_{*\dagger}(x_0, x)$	$p_{\diamond\diamond}(x_0, x)$	$p_{\diamond\dagger}(x_0, x)$
30.00	1.00000	0.00000	0.00000	1.00000	0.00000
35.00	0.98743	0.00354	0.00903	0.99097	0.00903
40.00	0.96998	0.00850	0.02152	0.97849	0.02152
45.00	0.94457	0.01620	0.03923	0.96077	0.03923
50.00	0.90624	0.02903	0.06474	0.93526	0.06474
55.00	0.84725	0.05106	0.10169	0.89831	0.10169
60.00	0.75677	0.08832	0.15491	0.84509	0.15491
65.00	0.62287	0.14700	0.23013	0.76987	0.23013

$$p_{\diamond\diamond}(x, y) = \exp\left(-\int_x^y \mu(\tau)d\tau\right),$$

which solve Kolmogorov's differential equations for this model:

$$\frac{d}{dt}p_{**}(s, t) = -p_{**}(s, t) \times (\mu(t) + \sigma(t)),$$

$$\frac{d}{dt}p_{*\diamond}(s, t) = -p_{*\diamond}(s, t)\,\mu(t) + p_{**}(s, t)\,\sigma(t),$$

$$\frac{d}{dt}p_{*\dagger}(s, t) = (p_{**}(s, t) + p_{*\diamond}(s, t)) \times \mu(t),$$

$$\frac{d}{dt}p_{\diamond *}(s, t) = 0,$$

$$\frac{d}{dt}p_{\diamond\diamond}(s, t) = -p_{\diamond\diamond}(s, t)\,\mu(t),$$

$$\frac{d}{dt}p_{\diamond\dagger}(s, t) = p_{\diamond\diamond}(s, t)\,\mu(t),$$

$$\frac{d}{dt}p_{\dagger\dagger}(s, t) = 0,$$

with the boundary conditions $p_{ij}(s, s) = \delta_{ij}$. Note that, if one uses a model with a positive probability of becoming active again, one has to modify the first, second, forth and fifth equation. Obviously one can solve these equations with numerical methods, the solutions for the given example are listed in Table 2.1.

Exercise 2.4.3 Consider the above system of differential equations.

1. Find an exact solution.
2. Find a numerical approximation to the solution.

Chapter 3
Interest Rate

3.1 Introduction

An important part of every insurance contract is the underlying interest rate. The so called technical interest rate describes the interest which the insurer guarantees to the insured. It is a significant factor for the size of the premiums. If the technical interest rate is too low it yields inflated premiums, if it is too high it might yield to insolvency of the insurance company.

For the technical interest rate one can use a deterministic or a stochastic model. In the latter case the interest rate will be coupled to the bond market. In the following we are going to define the main parameters connected to the interest rate and present their relations.

3.2 Definitions

Example 3.2.1 Suppose we put 10,000 USD on a bank account on the first of January. If at the end of the year there are 10,500 USD on the account, then the underlying interest rate was 5%.

Definition 3.2.2 (*Interest rate*) We denote by i the *yearly interest rate*. Furthermore we assume, that it depends on time and write $i_t, t \geq 0$. If we use a stochastic model for the interest rate, then i is a stochastic process $(i_t(\omega))_{t \geq 0}$.

One should note that this definition is useful in particular for the discrete time model. Since in this case one would use the time intervals which are given by the discretisation. The calculation of the future capital based on an interest rate is done by

$$B_{t+1} = (1 + i_t) \times B_t.$$

M. Koller, *Stochastic Models in Life Insurance*, EAA Series,
DOI: 10.1007/978-3-642-28439-7_3, © Springer-Verlag Berlin Heidelberg 2012

Here B_t denotes the value of the account at time t. Often also the inverse of this relation is relevant. Thus one defines the discount rate.

Definition 3.2.3 (*Discount rate*) Let i_t be the interest rate in year t. Then

$$v_t = \frac{1}{1+i_t}$$

is the *discount rate* in year t.

The discount rate can be used to calculate the net present value (the present value of the future benefits). If the interest rate is a stochastic process the previous considerations lead to the following problem: suppose we are going to receive 1 USD in one year, what is its present value? Generally there are two possible ways to find an answer.

Valuation principle A: If the interest rate i is known, the present value X is

$$X = \frac{1}{1+i}$$

and thus its mean is

$$X_A = E\left[\frac{1}{1+i}\right].$$

Valuation principle B: If the interest rate is known, the value of the account at the end of the year is $X(1+i)$. Thus

$$1 = E\left[X(1+i)\right] = X \times E\left[1+i\right]$$

holds and the mean is

$$X_B = \frac{1}{E\left[1+i\right]}.$$

But in general X_A is not equal to X_B. To overcome this problem one needs to fix by an assumption the valuation principle of the interest rate: we are always going to determine the net present value by valuation principle A (cf. [Büh92]).

After solving this paradox situation one understands the importance of the discount rate. Obviously, the same problem also arises in continuous time. For models in continuous time we assume that the interest is also payed continuously such that

$$B_{t+s} = \exp\left(\int_t^{t+s} \delta(\xi)\,d\xi\right) \times B_t.$$

Definition 3.2.4 (*Interest intensity*) The *interest intensity* at time t is denoted by $\delta(t)$.

A yearly interest rate i yields

$$e^\delta = 1 + i$$

and thus

$$\delta = \ln(1 + i).$$

In continuous time the discount rate (from t to 0) is

$$v(t) = \exp\left(-\int_0^t \delta(\xi)\, d\xi\right).$$

Here the discount rate is modelled from t to 0 which is contrary to the discrete setting. The following relation holds:

$$v_t = \exp\left(-\int_t^{t+1} \delta(\xi)\, d\xi\right).$$

For a stochastic interest rate the interest intensity δ is also a stochastic process $(\delta_t(\omega))_{t \geq 0}$. Also in the continuous time setting we are going to use valuation principle A.

Finally it is worth to note the difference between $v(t)$ which denotes the discounting back to time 0 and v_t which denotes the discount from time $t + 1$ to t. This later quantity is commonly used in actuarial sciences for recursion formulae.

3.3 Models for the Interest Rate Process

Now we want to describe the stochastic behaviour of the interest rate. We start with an analysis of the average interest rate of Swiss government bonds in Swiss franc. Figure 3.1 shows the rate from 1948 up to 2009.

The figure shows that during the given period the interest rate of bonds was subject to huge fluctuations. The minimal rate of 1.93% was reached in 2005. The maximum of 7.41% was recorded during the "oil crisis" in the autumn of 1974. It is interesting to note, that in the first printing of this book (1999) the minimum was actually reached in 1954 at about 2.5%. At this time nobody expected that the interest rates in Switzerland would drop that far again. Based on this observation the inherent risk of the technical interest rate in insurance policies becomes obvious.

After seeing these values one starts to wonder how the technical interest rate should be determined. First of all, this depends on the purpose of the model in use. One has to differentiate between short term and long term relations. Moreover, the interest rate might only be used for marketing or forecasting purposes. In any case

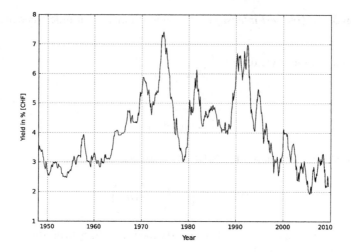

Fig. 3.1 Average yield of government bonds in %

it should be emphasised that it might be risky to fix a constant interest rate above the level of the observed minimum. Since in this case there might be periods during which the interest yield of the assets does not cover the liabilities. Thus one has to take care when fixing the technical interest rate.

Another method to determine the technical interest rate is based on the analysis of the yield-curve or the forward-curve. This curves allow to measure the interest rate structure. The yield-curve can be used to determine the interest rate which one would get for a bond with a fixed investment period. Figure 3.2 shows the yield-curve for various currencies, it indicates that the interest rate is smaller for a bond with a short investment period than for a bond with a long investment period. This is called a 'normal' interest rate structure. Conversely, one speaks of an 'inverse' interest rate structure if the bonds with a short investment period provide a better yield than those with a long investment period.

This point of view provides a realistic evaluation of the interest rate. To utilise this we are going to define the so called zero coupon bond.

Definition 3.3.1 (*Zero coupon bond*) Let $t \in \mathbb{R}$. Then the *zero coupon bond* with contract period t is defined by

$$\mathcal{Z}_{(t)} = (\delta_{t,\tau})_{\tau \in \mathbb{R}^+}.$$

Thus the zero coupon bound is a security, which has the value 1 at time t.

Definition 3.3.2 (*Price of a zero coupon bond*) Let $t \in \mathbb{R}$. Then the *price* of a zero coupon bond $\mathcal{Z}_{(s)}$ at time t with contract period s is denoted by

$$\pi_t \left(\mathcal{Z}_{(s)} \right).$$

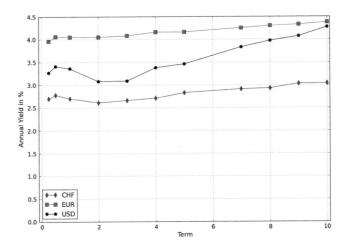

Fig. 3.2 Yield curves as at 1.1.2008

Based on these curves one can calculate the forward rates, i.e. the interest rate for year $n \rightsquigarrow n + 1$, of the corresponding investment:

$$(1 + i_k) = \frac{\pi_t \left(\mathcal{Z}_{(k)} \right)}{\pi_t \left(\mathcal{Z}_{(k+1)} \right)}.$$

Here i_k is called the forward rate at time t for the contract period $[k, k + 1[$, it is the *expected* rate for this time interval. Therefore the discount rate is given by

$$v_k = \frac{\pi_t \left(\mathcal{Z}_{(k+1)} \right)}{\pi_t \left(\mathcal{Z}_{(k)} \right)}.$$

Thus one can use a time dependent technical interest rate and adapt the necessary elements in the expectation based on the liabilities. This is especially useful for short term contracts with cash flows which are accessible. Consider for example the acquisition of a pension portfolio. In this case one takes over the obligation to pay the pensions of a given pension fund. The described method reduces the risks taken by fixing the technical interest rate.

The third method to determine the technical interest rate uses a stochastic interest rate model. This is interesting for practical and theoretical purposes. On the one hand this method is useful for policies which are tied to the performance of funds and which provide guaranties. On the other hand it enables us to derive models which provide a tool to measure the risk of an insurance portfolio with respect to changes of the interest rate (see Chap. 9). It turns out that for models with stochastic interest rate the corresponding risk does not vanish when the number of policies increases. This is a major difference to the deterministic model. Furthermore it means that the risk induced by the interest rate has a systemic, and thus dangerous, component.

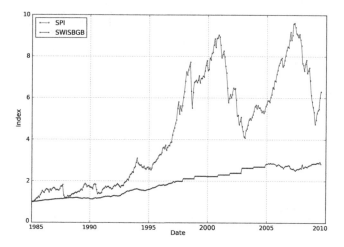

Fig. 3.3 Performance of different indices

The construction of these models requires an analysis of the returns of several investment categories. Figure 3.3 shows the performance of two indices. These are mea-
surements of the mean value of the return on investment in a given category. The
indices in Fig. 3.3 are the following

SPI   Swiss Performance Index: Swiss shares
SWISBGB   Swiss government bonds

Observing the performances of these indices (Fig. 3.3) we find a significant
difference between shares and bonds. The expected return is larger for shares, but
they also have a larger volatility (greater variance).

For a model with stochastic interest rate one has to model processes as depicted in
the Figs. 3.1, 3.2 and 3.3. The main difficulty in this context is the fact that there is no
common standard model. Thus the actuary is responsible for selecting an appropriate
model for the given problem.

In the following section we will describe some popular models. The reader interested in further details on the financial market, in particular interest rate models, is
referred for example to [Hul97].

3.4 Stochastic Interest Rate

In the previous section we have seen several possibilities to determine the value
of a cash flow based on the interest rate. Furthermore, the basics of a stochastic
interest rate model were introduced. In this section we will present explicitly several
stochastic interest rate models.

First of all one has to understand the difference between the stochastic behaviour of the interest rate and the stochastic component of the mortality. Both create a risk for an insurance company. On the one hand there is the risk induced by the fluctuation of the interest rate and on the other hand there is risk based on the individual mortality. Changes of the interest rate affect all policies to the same degree. But the variation of the risk based on the individual mortality decreases when the number of policies increases. This is due to the law of large numbers and the independence of individual lifetimes.

Now we give a brief survey of stochastic interest rate models. We will concentrate on a description of these models without rating them. Nevertheless one should note that some models (e.g. the random walk model) are unsuitable to realistically describe an interest rate process.

3.4.1 Discrete Time Interest Rate Models

Random walk: Let $\mu \in \mathbb{R}_+$ and $t \geq 0$. The interest rate i_t is defined by

$$i_t = \mu + X_t, \mu \in \mathbb{R},$$
$$X_t = X_{t-1} + Y_t,$$
$$Y_t \sim \mathcal{N}(0, \sigma^2) \text{ i.i.d.}$$

Note that this model is too simple to capture the real behaviour.

AR(1)-model: In this model the interest rate is an auto-regressive process of order 1:

$$i_t = \mu + X_t,$$
$$X_t = \phi X_{t-1} + Y_t, \text{ with } |\phi| < 1,$$
$$Y_t \sim \mathcal{N}(0, \sigma^2).$$

The model is mostly used by actuaries in England. The main idea is to start with an AR(1)-process as a model for the inflation. Then, in a second step, the models of the other economic values are based on this inflation. The constructed models have many parameters and are difficult to fit. References: [BP80, Wil86, Wil95].

3.4.2 Continuous Time Interest Rate Models

Brownian motion: $\delta_t = \delta + \sigma W_t$, where W_t is a standard Brownian motion.

Vasiček-model: The interest intensity is defined by the following stochastic differential equation

$$d\delta_t = -\alpha (\delta_t - \delta) \, dt + \sigma \, dW_t.$$

References: [Vas77]

Cox–Ingersoll–Ross: The interest intensity is defined by the following stochastic differential equation

$$d\delta_t = -\alpha\,(\delta_t - \delta)\,dt + \sigma\,\sqrt{\delta_t}\,dW_t.$$

References: [CIR85]

Markovian interest intensities: This model ([Nor95b]) uses a Markov chain $(X_t)_{t\geq 0}$ on a finite state space and a deterministic function $\delta_j(t)$ for each $j \in S$. Then the interest intensity is defined by

$$\delta_t = \sum_{j\in S} \chi_{\{X_t=j\}}\delta_j(t).$$

This means, that the interest intensity in a given state j at time t is determined by the corresponding deterministic function $\delta_j(.)$ evaluated at t. The model has the advantage, that it can be integrated into the Markov model. Furthermore it is very flexible due to its general state space. Therefore we will focus on this model in the following.

One should note that the Vasiček-model and the CIR-model feature mean reversion. Thus the interest intensity without stochastic noise (dW) converges in the long run toward the mean intensity δ, since the differential equation without stochastic noise

$$d\delta_t = -\alpha\,(\delta_t - \delta)\,dt$$

has the solution

$$\delta_t = \gamma \times \exp(-\alpha t) + \delta.$$

The Vasiček-model and the Cox–Ingersoll–Ross-model are often used to model interest rate processes in applications. We will use these models in Chap. 9.

Brownian motion and the Vasiček-model are problematic, since they allow negative interest rates with positive probability. In the Cox–Ingersoll–Ross model this can be prevented by an appropriate choice of the parameters.

In the following we assume that the presented stochastic differential equations have a solution.

We have seen above various models which are based on fundamentally different ideas. But in addition to the risk in the choice of the model there are further relevant systemic risks which affect the interest rate. These are:

The interest rate paid on an investment is not purely random. It also depends on political decisions. For example a monetary union causes the interest rates to converge, since in this case only one currency with one (random) interest rate exists (e.g. the European monetary union).

Chapter 4
Cash Flows and the Mathematical Reserve

4.1 Introduction

In the previous two chapters we introduced several types of insurances and their setup. Based on this we will now answer several fundamental questions.

First of all we will decide which general model we are going to use. Afterwards we will explain how to value and price an insurance policy.

The present value of an insurance policy, the so called mathematical reserve, has to be determined by an insurance company on a yearly basis for the annual statement. This is necessary since the company has to reserve this value. The mathematical reserve is also important for the insured when he wants to cancel his policy before maturity.

In the remainder of the chapter the insurance model from Chap. 1 will be combined with the stochastic models of Chap. 2. Obviously Markov chains with a countable state space are not the only possible stochastic model, but we will focus on these. On the one hand they are general enough to model many phenomena. On the other hand the corresponding formulas are simple enough to perform explicit calculations.

4.2 Examples

In this section we present some examples which motivate the use of the Markov chain model for insurance policies.

Example 4.2.1 (Life insurance) Usually the state space of a permanent life insurance consists of the states "dead" and "alive". Thus we use for the policy setup and for the stochastic process the state space $S = \{*, \dagger\}$, where $*$ denotes "alive" and $\dagger$ denotes "dead". Based on the benefits of such a policy, as described in Chap. 1, one has to model the corresponding stochastic process. We will use the exemplary life insurance from Chap. 2. A typical sample path of the stochastic process corresponding to this

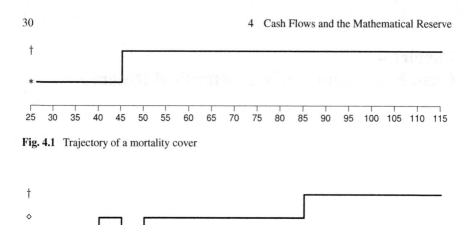

Fig. 4.1 Trajectory of a mortality cover

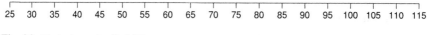

Fig. 4.2 Trajectory of a disability cover

policy is shown in Fig. 4.1. It indicates that at the time of death (here at $x = 45$) the corresponding payment (e.g. 200,000 USD) is due. The mortality at that time is:

$$\mu_{*\dagger}(x)|_{x=45} = \exp(-9.13275 + 0.08094x - 0.000011x^2)|_{x=45}$$
$$= 0.00404.$$

This means that on average 4 out of 1000 forty-five year old men die per year.

Up to now we are not able to calculate the premiums for the policy in the example above. But we already notice the interplay between the payments and the stochastic processes.

Example 4.2.2 (Temporary disability pension) In this example we consider a policy of a disability pension which corresponds to the sample path in Fig. 4.2. We want to record the various cash flows which it induces. For this we take the transition intensities from Example 2.4.2 with the additional assumption $\mu_{\diamond*}(x) = 0.05$. Then the sample path presented in Fig. 4.2 causes the cash flows listed in Table 4.1.

4.3 Fundamentals

In order to derive realistic models we have to know the fundamentals (underlying probabilities and biometric quantities). They are especially needed for the calculation of the premiums and mathematical reserves. As actuary one can look up the fundamentals in published tables. These list the probability of given events, for example the probability to die at a given age.

Table 4.1 Example of cash flows for a disability pension

Time	State	Cash flow	μ
$x \in [0, 40[$	Active ($*$)	Premiums	
$x = 40$	Becomes disabled	Disability capital	$\mu_{*\diamond} = 0.00214$
$x \in]40, 45[$	Disabled ($\diamond$)	Disability pension	
$x = 45$	Becomes active	–	$\mu_{\diamond*} = 0.05000$
$x \in [45, 50[$	Active ($*$)	Premiums	
$x = 50$	Becomes disabled	(maybe) Disability capital	$\mu_{*\diamond} = 0.00387$
$x \in]50, 85[$	Disabled ($\diamond$)	Disability pension	
From 65		Pension	
$x = 85$	Dead	Sum payable at death	$\mu_{\diamond\dagger} = 0.12932$

Table 4.2 Average life expectancy based on Swiss mortality tables (male)

Alter	1881–1888	1921–1930	1939–1944	1958–1963	1978–1983	1988–1993	1998–1903
1	51.8	61.3	64.8	69.4	72.1	73.8	76.6
20	39.6	45.2	47.9	51.5	53.8	55.3	58.0
40	25.1	28.3	30.4	32.8	35.1	36.8	39.0
60	12.4	13.8	14.8	16.2	17.9	19.3	21.1
75	5.6	6.2	6.6	7.5	8.5	9.2	10.3

Table 4.3 Average life expectancy based on Swiss mortality tables (female)

Alter	1881–1888	1921–1930	1939–1944	1958–1963	1978–1983	1988–1993	1998–1903
1	52.8	63.8	68.5	74.5	78.6	80.5	82.2
20	41.0	47.6	51.3	56.2	60.1	61.8	63.4
40	26.7	30.9	33.4	37.0	40.7	42.5	43.8
60	12.7	15.1	16.7	19.2	22.4	24.0	25.2
75	5.7	6.7	7.4	8.6	10.7	11.9	12.8

The tables used by insurance companies often incorporate a certain spread. For example one increases the probability of dying at a certain age if one calculates a life insurance. Conversely, one increases the survival rate if one calculates a pension. This spread is used to decrease the risk of default and to cover possible demographic trends. That this is necessary illustrate for example the Tables 4.2 and 4.3. They list the average life expectancy for several generations, as given in the Swiss mortality tables. The average life expectancy is the number of years which a person of given age has on average still ahead. These tables clearly show that the average life expectancy increased during the last one hundred years. Therefore a spread is clearly necessary to cover this trend. Other western countries experience a similar increase of the average life expectancy.

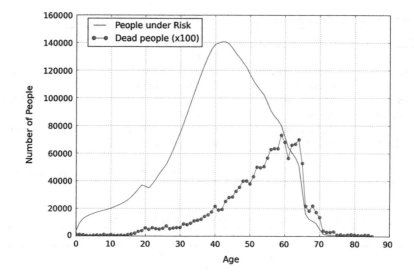

Fig. 4.3 Inforce and number of dead people

We note that demographic quantities are constantly changing, like the average life expectancy. But where does this data actually come from and how where these tables calculated?

For the mortality tables one uses either the samples which are owned by the given insurance company or a collection of samples which is obtained jointly by several insurers. Then, to calculate the mortality rate, one counts the number of persons at risk and the number of died subjects for a given period of time (e.g. five years). The following example is based on data obtained by a large Swiss insurance company [PT93]. Figure 4.3 shows the number of alive and dead people at a given age. Figure 4.4 shows the raw mortality and the smoothed mortality.

The smoothed mortality is obtained by a smoothing algorithm. We are not going to discuss these algorithms. But we want to note, that there are various algorithms which greatly differ in their complexity.

On the raw curve one notices for example the accident-bump (i.e. the increased mortality) between 15 and 25 years. This is not visible in the smooth curve. Thus one has to adjust in this region the smoothed mortality.

In this example a polynomial of degree two was fitted to $\log(\mu_{*\dagger})$:

$$\mu_{*\dagger}(x) = \exp(-7.85785 + 0.01538 \cdot x + 5.77355 \cdot 10^{-4} \cdot x^2).$$

Analogous other demographic quantities which are relevant for the calculation of the premiums are obtained. Also for these a smoothed curve is obtained by an application a smoothing algorithm to the raw data.

Table 4.4 Typical quantities for the calculation of premiums

Variable	Meaning
q_x	Mortality, possibly separate for accidents and illness,
i_x	Probability of becoming disabled, possibly separate for accidents and illness,
r_x	Probability of becoming active again, possibly partitioned by the lengths of the disability period,
g_x	Average degree of disability,
h_x	Probability of being married at the time of death,
y_x	Average age of the surviving marriage partner at the death of the insured.

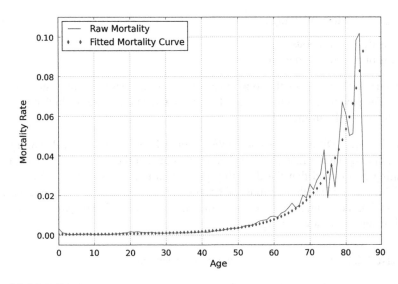

Fig. 4.4 Mortality man

The relevant probabilities and biometric quantities are collected in a catalogue. Then, with the help of such a catalogue, one can calculate the premiums and values of various products.

Table 4.4 lists the typically relevant quantities.

Further details about the calculation of mortality tables and disability tables for the German and European market can be found in [DAV09].

4.4 Deterministic Cash Flows

Definition 4.4.1 (*Payout function*) A *deterministic payout function* A is a function

$$A : T \to \mathbb{R}, t \mapsto A(t),$$

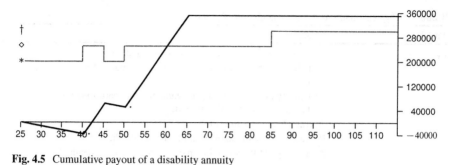

Fig. 4.5 Cumulative payout of a disability annuity

on $T \subset \mathbb{R}$ with the following properties:

1. A is right continuous,
2. A is of bounded variation.

The value $A(t)$ represents the total payments from the insurer to the insured up to time t. The payout functions are those functions in the policy setup which represent the benefits for the insured.

Example 4.4.2 (Disability insurance) We continue Example 4.2.2 by calculating the corresponding payout function. For this we assume that the policy does not contain a waiting period and that the disability pension is fixed to 20,000 USD per year (until the age of 65) with premiums of 2,500 USD per year until 65. Furthermore we suppose that the insurance was contracted at the age $x_0 = 25$. The payout function of this example is shown in Fig. 4.5.

Exercise 4.4.3 Derive the payout function for Example 4.4.2.

Remark 4.4.4 (Functions of bounded variation) Functions of bounded variation have the following properties [DS57]:

1. A function A of bounded variation corresponds to a measure on $\sigma(\mathbb{R})$. We will denote the measure again by A. This measure is called *Stieltjes measure*. In our setting it is also called payout measure.
2. Let A be a function of bounded variation on $\mathbb{R}$. Then there exist two positive, increasing and bounded functions B and C such that $A = B - C$. In the insurance model we can interpret B as inflow and C as outflow of cash. This representation is unique if one assumes that the measures corresponding to B and C have disjoint support. (Exercise: Calculate B and C for Example 4.4.2.)
3. Let A be the measure corresponding to a function of bounded variation on $\mathbb{R}$. Then A can be decomposed uniquely into a discrete measure μ and a continuous measure ψ. Furthermore one can decompose ψ into a part which is absolute continuous with respect to the Lebesgue measure and a remainder part. The support of μ is a countable set since A is finite on bounded sets.

4. Let A be a function of bounded variation and $T \in \sigma(\mathbb{R})$. Then $A \times \chi_T$ is also a function of bounded variation. (Here the function χ_T is the indicator function introduced in Definition 2.1.2.)

The above properties also hold for payout functions, since these are just functions of bounded variation. The decomposition of the Stieltjes measures will be useful later on. Thus we introduce the following notation:

Definition 4.4.5 (*Decomposition of measures*) Let f be a function of bounded variation with corresponding Stieltjes measure A. Then we define

$$\mu_f := A.$$

We know that we can decompose this measure uniquely into $A = B - C$, where B and C are positive measures with disjoint support. Therefore we define:

$$A^+ := B,$$
$$A^- := C.$$

Furthermore the Stieltjes measure A can be decomposed uniquely into $A = D + E$, where D is discrete and E is continuous. Therefore we define:

$$A^{\text{atom}} := D,$$
$$A^{\text{cont}} := E.$$

Furthermore, let μ be a measure which is absolute continuous with respect to the Lebesgue measure λ. Then we denote by $\frac{d\mu}{d\lambda}$ the *Radon-Nikodym density* of μ with respect to λ.

Above we have seen the most important properties of deterministic cash flows. This enables us to define their values with the help of the discount rate. Recall, the discount rate is given by

$$v(t) = \exp\left(-\int_0^t \delta(\tau)\,d\tau\right).$$

Then the present value of a cash flow is defined as follows.

Definition 4.4.6 (*Value of a cash flow*) Let A be a deterministic cash flow and $t \in \mathbb{R}$. We define:

1. The value of a cash flow A at time t is

$$V(t, A) := \frac{1}{v(t)} \int_0^\infty v(\tau)\,dA(\tau).$$

2. The value of the future cash flow is

$$V^+(t, A) := V(t, A \times \chi_{]t,\infty]}).$$

It is also called *prospective value of the cash flow* or *prospective reserve*.

Concerning these definitions one should note:

Remark 4.4.7   1. The idea of the prospective reserve is to calculate the present
value of the future cash flows. Thus a payment of ζ which is due in two years
contributes $v(2) \times \zeta$ to the present value. Initially the reserves are defined for
deterministic cash flows. To define them also for random cash flows one uses
the corresponding conditional expectations.
2. The definition implicitly requires that $v(t)$ is integrable with respect to the measure A, i.e. $v \in L^1(A)$.
3. The equation $A = A^{\text{atom}} + A^{\text{cont}}$ also implies $V(t, A) = V(t, A^{\text{atom}}) + V(t, A^{\text{cont}})$. This decomposition allows us to use different methods of proof
for the discrete and the continuous part of the measure.

Example 4.4.8   We want to calculate $V^+(t, A)$ for the cash flow defined in Example
4.4.2 with $\delta(\tau) = \log(1.04)$. The first step is to calculate A^+ and A^-:

$$dA^+ = 20000 \, (\chi_{[40,45[} + \chi_{[50,65[})d\tau,$$
$$dA^- = 2500 \, (\chi_{[25,40[} + \chi_{[45,50[})d\tau.$$

Then we get for $t \in [25, 65[$

$$V^+(t, A) = 20000 \int_t^{65} (1.04)^{-(\tau-t)} \, (\chi_{[40,45[} + \chi_{[50,65[})d\tau$$
$$-2500 \int_t^{65} (1.04)^{-(\tau-t)} \, (\chi_{[25,40[} + \chi_{[45,50[})d\tau.$$

4.5 Stochastic Cash Flows

Definition 4.5.1 (*Stochastic cash flow*) A *stochastic cash flow* or a *stochastic process
of bounded variation* is a stochastic process $(X_t)_{t \in T}$ for which almost all sample paths
are functions of bounded variation.

Let A be a stochastic process of bounded variation such that $t \mapsto A_t(\omega)$ is right
continuous and increasing for each $\omega \in \Omega$. Then it is possible to calculate the integral
$\int f(\tau)d\mu_{A.(\omega)}(\tau)$ for a bounded Borel function f. Similarly, one can define P-almost
everywhere the integral $\int f(\tau, \omega)d\mu_{A.(\omega)}(\tau)$ if $F_t = f(t, \omega)$ is a bounded function
which is measurable with respect to the product sigma algebra. The construction
of these integrals can be extended to general processes of bounded variation by

decomposing the sample path, a function of bounded variation, into its positive (increasing) and negative (decreasing) part.

Definition 4.5.2 Let $(A_t)_{t \in T}$ be a process of bounded variation on $(\Omega, \mathcal{A}, P)$ and $F : \mathbb{R} \times \Omega \to \mathbb{R}$ be a bounded and product measurable function. Then the description above yields the following definition

$$(F \cdot A)_t(\omega) = \int_0^t F(\tau, \omega) dA_\tau(\omega).$$

We also write this relation in the symbolic notation of stochastic differential equations:

$$d(F \cdot A) = F \, dA.$$

This definition allows us to give a precise definition of the stochastic cash flows in our insurance model.

Definition 4.5.3 (*Policy cash flows*) We consider an insurance policy with state space S and payout functions $a_{ij}(t)$ and $a_i(t)$.[1] Based on Definition 2.1.8 we can define the stochastic cash flows corresponding to an insurance policy by

$$dA_{ij}(t, \omega) = a_{ij}(t) \, dN_{ij}(t, \omega),$$
$$dA_i(t, \omega) = I_i(t, \omega) \, da_i(t),$$
$$dA = \sum_{i \in S} dA_i + \sum_{(i,j) \in S \times S, i \neq j} dA_{ij}.$$

The quantity $A_{ij}(t, \omega)$ is the sum of the random cash flows which are induced by transitions from state i to state j up to time t. Similarly, $A_i(t, \omega)$ represents the sum of the random cash flows up to time t which are pension payments for being in state i.

Remark 4.5.4   1. The quantity $dA_{ij}(t, \omega)$ corresponds to the increase of the liabilities by a transition $i \rightsquigarrow j$. Therefore $A_{ij}(t, \omega)$ increases at time t by the capital benefit $a_{ij}(t)$ if at time t a transition $i \rightsquigarrow j$ takes place, i.e. if $N_{ij}(t)$ increases by 1. Similarly, $dA_i(t)$ corresponds to the increase of the liabilities caused by the insured being in state i.

2. The integrals appearing above are well defined since the corresponding processes are, by definition, of bounded variation. Moreover also the payout functions have the required regularity.

3. The quantities $(F \cdot A)_t$ are measurable for each t since F was assumed to be product measurable. Therefore also the expectation $E[(F \cdot A)_t]$ is well defined. Similarly, the conditional expectations $E[(F \cdot A)_t \mid \mathcal{F}_s]$ are well defined.

4. Thus one can apply Definition 4.4.6 (value of a cash flow) point wise (i.e. for each sample) to a stochastic cash flow. This yields the equation

---

[1] Thus the functions are of bounded variation and, in particular, bounded.

$$dV(t, A) = v(t)\, dA(t)$$

$$= v(t) \left[\sum_{i \in S} I_i(t)\, da_i(t) + \sum_{(i,j) \in S \times S,\, i \neq j} a_{ij}(t)\, dN_{ij}(t) \right].$$

5. In the discrete Markov model at most two cash flows occur during a time interval $[t, t+1[$. Firstly, if the policy is in state i, $a_i^{\text{Pre}}(t)$ is paid at the beginning of the interval. Secondly, if there is a transition $i \rightsquigarrow j$, $a_{ij}^{\text{Post}}(t)$ is due at the end of the interval. Hence the following equations can be used to calculate the total cash flows:

$$\Delta A_{ij}(t, \omega) = \Delta N_{ij}(t, \omega) a_{ij}^{\text{Post}}(t), \tag{4.1}$$

$$\Delta A_i(t, \omega) = I_i(t, \omega) a_i^{\text{Pre}}(t), \tag{4.2}$$

$$\Delta A(t, \omega) = \sum_{i \in S} \Delta A_i(t, \omega) + \sum_{i,j \in S} \Delta A_{ij}(t, \omega), \tag{4.3}$$

where $\Delta A(t)$ (and similarly ΔA_{ij}, and ΔA_i, respectively) stands for the change of $A(t)$ from t to $t+1$, e.g. $\Delta A(t) := A(t+1) - A(t)$.

Definition 4.5.5 Let A and (optionally) also v be stochastic processes on $(\Omega, \mathcal{A}, P)$ which are adapted to the filtration $\mathbb{F} = (\mathcal{F}_t)_{t \geq 0}$. In this case the *prospective reserve* is defined by:

$$V_{\mathbb{F}}^+(t, A) = E\left[V^+(t, A) \mid \mathcal{F}_t\right].$$

One should note that also these reserves, like the usual expectations, might not exist, i.e., they might be infinite. In the following we will always assume that $V_{\mathbb{F}}^+(t, A)$ and the other quantities exist. This assumption is always satisfied in applications.

For a Markov chain the conditional expectation with respect to $\mathcal{F}_t$ depends only on the state at time t. Thus we additionally define

$$V_j^+(t, A) = E\left[V^+(t, A) \mid X_t = j\right].$$

The following definition fixes our assumptions on the regularity of an insurance model.

Definition 4.5.6 (*Regular insurance model*) A regular insurance model consists of:

1. a regular Markov chain $(X_t)_{t \in T}$ with a state space S,
2. payout functions $a_{ij}(t)$ and $a_i(t)$,
3. right continuous interest intensities $\delta_i(t)$ of bounded variation.

4.6 Mathematical Reserve

The mathematical reserve is the amount of money an insurance company has to reserve for the expected liabilities in order to remain solvent. We assume that the interest intensity δ has the following structure: $\delta_t = \sum_{j \in S} I_j(t)\, \delta_j(t)$. Then the required reserves for the cash flows are defined by:

Definition 4.6.1 (*Mathematical reserve*) The mathematical reserve for being in state $g \in S$ within a time interval $T \in \sigma(\mathbb{R})$ under the condition $X_t = j$ is defined by

$$V_j(t, A_{gT}) = E\left[\frac{1}{v(t)} \times \int_T v(\tau)\, dA_g(\tau) \mid X_t = j\right].$$

Similarly, for transitions from g to $h \in S$, we define

$$V_j(t, A_{ghT}) = E\left[\frac{1}{v(t)} \times \int_T v(\tau)\, dA_{gh}(\tau) \mid X_t = j\right].$$

We use the notation $V_j(t, A_g)$ and $V_j(t, A_{gh})$ for $V_j(t, A_{g\mathbb{R}})$ and $V_j(t, A_{gh\mathbb{R}})$, respectively.

Remark 4.6.2 The definitions of the mathematical reserve can be translated to the discrete model. One just has to replace the integrals by the corresponding sums:

$$V_j(t, A_{gT}) = E\left[\frac{1}{v(t)} \times \sum_{\tau \in T} v(\tau)\, \Delta A_g(\tau) \mid X_t = j\right].$$

Analogous, for transitions from g to $h \in S$ we set:

$$V_j(t, A_{ghT}) = E\left[\frac{1}{v(t)} \times \sum_{\tau \in T} v(\tau + 1)\, \Delta A_{gh}(\tau) \mid X_t = j\right],$$

where we assumed that the payments always take place at time $\tau + 1$.

Therefore the total reserve (or mathematical reserve) for a given state j is

$$V_j(t, A) = \sum_{g \in S} V_j(t, A_g) + \sum_{g,h \in S, g \neq h} V_j(t, A_{gh})$$

for the continuous time model and

$$V_j(t, A) = \sum_{g \in S} V_j(t, A_g^{\mathrm{Pre}}) + \sum_{g,h \in S} V_j(t, A_{gh}^{\mathrm{Post}})$$

for the discrete time model. Thus we have defined the mathematical reserves. The next step is to calculate their values. Let us consider the relevant cash flows. On the one hand there are flows of the form $dA_1(t) = a(t)dN_{jk}(t)$ and on the other hand are flows of the form $dA_2(t) = I_j(t)dA(t)$.

The first step is to calculate the integrals $\int dA$ for the partial cash flows. Afterwards we will derive explicit formulas for the mathematical reserves.

Theorem 4.6.3 *Let $(X_t)_{t \in T}$ be a regular Markov chain on $(\Omega, \mathcal{A}, P)$ (cf. Def. 2.3.2). Furthermore let $i, j, k \in S$, $s < t$ and $T \in \sigma(\mathbb{R})$ where $T \subset [s, \infty]$. Then the following statements hold:*

1.

$$E\left[\int_T a(\tau)\, dN_{jk}(\tau) \mid X_s = i\right] = \int_T a(\tau)\, p_{ij}(s, \tau)\, \mu_{jk}(\tau)d\tau$$

for $a \in L^1(\mathbb{R})$.
2. *Let A be a function of bounded variation, then*

$$E\left[\int_T I_j(\tau)\, dA(\tau) \mid X_s = i\right] = \int_T p_{ij}(s, \tau)\, dA(\tau).$$

Proof   1.   The step functions are dense in L^1. Therefore it is enough to show the equality for functions of the form $\chi_{[a,b]}$. Moreover, the Borel σ-algebra is generated by the intervals in $\mathbb{R}_+$. Thus we can take $T = [c, d]$. Further we can set $c = a$ and $d = b$ without loss of generality, since the indicator function is equal to zero outside the interval $[a, b]$.
Define the function

$$h(t) := E\left[N_{jk}(t) \mid X_s = i\right].$$

Based on this definition we get

$$
\begin{aligned}
h(t + \Delta t) - h(t) &= E\left[N_{jk}(t + \Delta t) - N_{jk}(t) \mid X_s = i\right] \\
&= \sum_{l \in S} E\left[\chi_{\{X_t = l\}}(N_{jk}(t + \Delta t) - N_{jk}(t)) \mid X_s = i\right] \\
&= \sum_{l \in S} E\left[N_{jk}(t + \Delta t) - N_{jk}(t) \mid X_t = l\right] \times p_{il}(s, t).
\end{aligned}
$$

Now we observe that all the terms where $j \neq l$ are of order $o(\Delta t)$. Thus we get

$$= p_{ij}(s, t) \times \mu_{jk}(t) \times \Delta t + o(\Delta t).$$

Therefore $h'(t) = p_{ij}(s, t)\, \mu_{jk}(t)$, and an integration of this equation with initial condition $h(0) = 0$ yields the first statement of the theorem.
2.   For the second statement one has to interchange the order of integration, which is allowed by Fubini's theorem.

Remark 4.6.4 Also these statements can be translated easily to the discrete model. One gets the equations

$$E\left[\sum_{\tau \in T} a(\tau)\,\Delta N_{jk}(\tau)\mid X_s = i\right] = \sum_{\tau \in T} a(\tau)\,p_{ij}(s,\tau)p_{jk}(\tau,\tau+1)$$

and

$$E\left[\sum_{\tau \in T} I_j(\tau)\,\Delta A(\tau)\mid X_s = i\right] = \sum_{\tau \in T} p_{ij}(s,\tau)\,\Delta A(\tau).$$

Exercise 4.6.5 Complete the proof of Theorem 4.6.3.

An important consequence of Theorem 4.6.3 is the next theorem.

Theorem 4.6.6 *Let the assumptions of Theorem 4.6.3 be satisfied. Then*

$$dM_{ij}(t) := dN_{ij}(t) - I_i(t)\,\mu_{ij}(t)dt$$

is a martingale.

Proof We have

$$N_{ij}(t) \in L^1(\Omega, \mathcal{A}, P)$$

and

$$\int_0^t I_j(\tau)\,\mu_{ij}(\tau)d\tau \in L^1(\Omega, \mathcal{A}, P),$$

which implies

$$M_{ij}(t) \in L^1(\Omega, \mathcal{A}, P).$$

Next, we have to prove the equality $E[M_{ij}(t)\mid \mathcal{F}_s] = M_{ij}(s)$ for $s < t$. But, since the processes M, N and I are all derived from $(X_t)_{t \in T}$, it is enough to prove $E[M_{ij}(t)\mid X_s = k] = M_{ij}(s)$. But this is true, since

$$E[M_{ij}(t)\mid X_s = k] - M_{ij}(s) = E\left[\int_s^t dM_{ij}(\tau)\mid X_s = k\right]$$
$$= E\left[\int_s^t dN_{ij}(t) - I_i(t)\,\mu_{ij}(t)dt \mid X_s = k\right]$$
$$= 0,$$

where we used Theorem 4.6.3 in the last step.

Another application of Theorem 4.6.3 yields the following equations for the mathematical reserves in our insurance model:

Theorem 4.6.7 *Let a_{ij} and a_i be payout functions and $(X_t)_{t\in T}$ be a regular Markov chain on $(\Omega, \mathcal{A}, P)$. Then the following equations hold for fixed interest intensities (i.e., $\delta_i = \delta$):*

$$E[V(t, A_{jT})|\, X_s = i]$$
$$= \frac{1}{v(t)} \int_T v(\tau)\, p_{ij}(s, \tau) da_j(\tau),$$
$$E[V(t, A_{jkT})|\, X_s = i]$$
$$= \frac{1}{v(t)} \int_T v(\tau)\, a_{jk}(\tau)\, p_{ij}(s, \tau)\, \mu_{jk}(\tau) d\tau,$$
$$E[V(t, A_{jS})V(t, A_{lT})|\, X_s = i]$$
$$= \frac{1}{v(t)^2} \int_{T\times S} v(\theta)v(\tau)\{\chi_{\{\theta\le\tau\}} p_{ij}(s, \theta) p_{jl}(\theta, \tau)$$
$$+ \chi_{\{\theta>\tau\}} p_{il}(s, \tau) p_{lj}(\tau, \theta)\} da_j(\theta) da_l(\tau),$$
$$E[V(t, A_{jkS})V(t, A_{lmT})|\, X_s = i]$$
$$= \frac{1}{v(t)^2}\left[\int\!\!\int_{T\times S} v(\theta)v(\tau)\{\chi_{\{\theta\le\tau\}} p_{ij}(s, \theta) p_{kl}(\theta, \tau)\right.$$
$$+ \chi_{\{\theta>\tau\}} p_{il}(s, \theta) p_{mj}(\theta, \tau)\}\mu_{jk}(\theta)\mu_{lm}(\tau) a_{jk}(\theta) a_{lm}(\tau) d\theta d\tau$$
$$\left.+ \delta_{jk,lm} \int_{T\cap S} v(\tau)^2 p_{ij}(s, \tau)\mu_{jk}(\tau) a_{jk}^2 d\tau\right],$$
$$E[V(t, A_{jS})V(t, A_{lmT})|\, X_s = i]$$
$$= \frac{1}{v(t)^2} \int_{T\times S} v(\theta)v(\tau)\{\chi_{\{\theta\le\tau\}} p_{ij}(s, \theta) p_{jl}(\theta, \tau)$$
$$+ \chi_{\{\theta>\tau\}} p_{il}(s, \tau) p_{mj}(\tau, \theta)\} da_j(\theta)\mu_{lm}(\tau) a_{lm}(\tau) d\tau.$$

Proof The first two equations are a direct consequence of Theorem 4.6.3. For a proof of the remaining equations we refer to [Nor91]. ∎

Remark 4.6.8 Also this theorem can easily be translated to the discrete setting. The following equalities hold:

$$E[V(t, A_{jT})|\, X_s = i] = \frac{1}{v(t)} \sum_{\tau\in T} v(\tau)\, p_{ij}(s, \tau) a_j^{\text{Pre}}(\tau),$$
$$E[V(t, A_{jkT})|\, X_s = i] = \frac{1}{v(t)} \sum_{\tau\in T} v(\tau+1)\, p_{ij}(s, \tau)\, p_{jk}(\tau, \tau+1)\, a_{jk}^{\text{Post}}(\tau),$$

where we used that, for a transition $j \rightsquigarrow k$, the payments $a_{jk}^{\text{Post}}(\tau)$ are made the end of the period.

Exercise 4.6.9 Complete the proof of Theorem 4.6.7.

Given the transition probabilities one can use Theorem 4.6.7 to calculate the expectations and variances of the prospective reserves for each cash flow. Then, based on these partial reserves, one can calculate the total prospective reserves by the following result.

Theorem 4.6.10 *Let a regular insurance model (Definition 4.5.6) with deterministic interest intensities be given. Then the prospective reserves are given by*

$$V_j^+(t) = \frac{1}{v(t)} \int_{]t,\infty[} v(\tau) \sum_{g \in S} p_{jg}(t, \tau)$$

$$\times \left\{ da_g(\tau) + \sum_{S \ni h \neq g} a_{gh}(\tau) \mu_{gh}(\tau) d\tau \right\}.$$

Remark 4.6.11 The formula of the previous theorem is not very useful, since one has to calculate integrals based on the transition probabilities p_{ij}. This becomes even more complicated by the fact that in applications often only the μ_{ij} are given. In the next section we will find a more elegant way to calculate this quantity.

4.7 Recursion Formulas for the Mathematical Reserves

In this section we will derive a recursion formula for the reserves based on their integral representation. This recursion can be used in two ways. On the one hand it can be used to prove Thiele's differential equation. On the other hand it can be applied to the discrete model. Thereby it provides a way to calculate the values for various types of insurances. We will see in the remaining sections that these recursion equations, difference equations and differential equations are a extremely useful tools for explicit calculations. We adapt our definition of mathematical reserves, in order to simplify the proofs:

Definition 4.7.1 We define for a regular insurance model (Definition 4.5.6):

$$W_j^+(t) := v(t) \, V_j^+(t).$$

The difference between W and the usual mathematical reserve V is only the discount factor. V resembles the value of the cash flow at time t, whereas W is the value of the cash flow at time 0. Thus W is only a new notation which will help to keep the proofs simple. Based on this we are now able to derive a recursion formula for the prospective reserve.

Lemma 4.7.2 *Let $j \in S$, $s < t < u$ and $(X_t)_{t \in T}$ be a regular insurance model in continuous time with deterministic interest intensities. Then the following equation holds:*

$$W_j^+(t) = \sum_{g \in S} p_{jg}(t,u) W_g^+(u)$$

$$+ \int_{]t,u]} v(\tau) \sum_{g \in S} p_{jg}(t,\tau) \left\{ da_g(\tau) + \sum_{S \ni h \neq g} a_{gh}(\tau)\mu_{gh}(\tau)d\tau \right\}.$$

Proof The proof is based on the Chapman-Kolmogorov equation. We get

$$W_j^+(t) = \int_{]t,\infty]} v(\tau) \sum_{g \in S} p_{jg}(t,\tau) \left\{ da_g(\tau) + \sum_{S \ni h \neq g} a_{gh}(\tau)\mu_{gh}(\tau)d\tau \right\}$$

$$= \left(\int_{]t,u]} + \int_{]u,\infty]} \right) v(\tau) \sum_{g \in S} p_{jg}(t,\tau)$$

$$\times \left\{ da_g(\tau) + \sum_{S \ni h \neq g} a_{gh}(\tau)\mu_{gh}(\tau)d\tau \right\}$$

$$= \int_{]t,u]} v(\tau) \sum_{g \in S} p_{jg}(t,\tau) \left\{ da_g(\tau) + \sum_{S \ni h \neq g} a_{gh}(\tau)\mu_{gh}(\tau)d\tau \right\}$$

$$+ \int_{]u,\infty]} v(\tau) \sum_{g \in S} \left(\sum_{k \in S} p_{jk}(t,u) p_{kg}(u,\tau) \right)$$

$$\times \left\{ da_g(\tau) + \sum_{S \ni h \neq g} a_{gh}(\tau)\mu_{gh}(\tau)d\tau \right\}$$

$$= \int_{]t,u]} v(\tau) \sum_{g \in S} p_{jg}(t,\tau) \left\{ da_g(\tau) + \sum_{S \ni h \neq g} a_{gh}(\tau)\mu_{gh}(\tau)d\tau \right\}$$

$$+ \sum_{k \in S} p_{jk}(t,u) \left(\int_{]u,\infty]} v(\tau) \sum_{g \in S} p_{kg}(u,\tau) \right.$$

$$\left. \times \left\{ da_g(\tau) + \sum_{S \ni h \neq g} a_{gh}(\tau)\mu_{gh}(\tau)d\tau \right\} \right)$$

$$= \sum_{g \in S} p_{jg}(t,u) W_g^+(u)$$

$$+ \int_{]t,u]} v(\tau) \sum_{g \in S} p_{jg}(t,\tau) \left\{ da_g(\tau) + \sum_{S \ni h \neq g} a_{gh}(\tau)\mu_{gh}(\tau)d\tau \right\}.$$

The recursion formula can also be translated to the discrete model. For this one assumes that the payments are done at discrete times rather than in continuous time. For example pensions are paid at the beginning of the interval and death capital is paid at the end of the interval. We denote the payments at the beginning of the year by $a_i^{Pre}(t)$ and those at the end of the year by $a_{ij}^{Post}(t)$. Thus, in particular we assume that a transition between states can only occur at the end of the year.

Setting $\Delta t = 1$ in the previous lemma yields the following recursion for the reserves in the discrete setting.

Theorem 4.7.3 (Thiele's difference equation) *For a discrete time Markov model the prospective reserve satisfies the following recursion:*

$$V_i^+(t) = a_i^{Pre}(t) + \sum_{j \in S} v_t \, p_{ij}(t)\{a_{ij}^{Post}(t) + V_j^+(t+1)\}.$$

Remark 4.7.4 • The formula shows that errors which are introduced by the discretisation of time are due to payments between the discretisation times.
• The recursion formula of the mathematical reserve is very important for applications, since it provides a way to calculate a single premium and yearly premiums. In fact it is the most important formula for explicit calculations.
• To solve a differential equation or a difference equation one needs a boundary condition. For example, if one calculates a pension, the boundary condition is given by the fact that the reserve has to be equal to zero at the final age ω.

4.8 Calculation of the Premiums

In this section we are going to calculate single premiums and yearly premiums for several types of insurance policies. The calculations in the examples are based on the discrete recursion (Theorem 4.7.3). We start with an endowment policy.

Example 4.8.1 (Endowment policy in discrete time) We consider the insurance defined in Example 4.2.1. Thus there is a death benefit of 200,000 USD. Moreover we assume an endowment of 100,000 USD and a starting age of 30 with 65 as fixed age at maturity.

• How much is a single premium for this insurance, given a technical interest rate of 3.5%?
• How much are the corresponding yearly premium?

We use the mortality rates given by (2.13). First we calculate the single premium. The following payout functions are given:

$$a_{*\dagger}^{Post}(x) = \begin{cases} 200000, & \text{if } x < 65, \\ 0, & \text{otherwise,} \end{cases}$$

$$a_{**}^{Post}(x) = \begin{cases} 0, & \text{if } x < 64, \\ 100000, & \text{if } x = 64, \\ 0, & \text{otherwise.} \end{cases}$$

An application of Theorem 4.7.3 yields the results presented in Table 4.5. Here one has to note that the mathematical reserves for the case of survival and the case of death

Table 4.5 Reserves for
an endowment

Age	q_x	Res. for endowment	Res. for death benefit	Sums reserve
65	0.01988	**100000**	**0**	**100000**
64	0.01836	94844	3548	98392
63	0.01696	90083	6647	96730
62	0.01566	85674	9348	95022
61	0.01446	81579	11696	93275
60	0.01336	77768	13730	91498
55	0.00897	62086	20275	82360
50	0.00602	50444	22766	73210
45	0.00404	41470	22956	64426
40	0.00271	34362	21874	56236
35	0.00181	28624	20135	48759
30	0.00121	23928	18116	42044

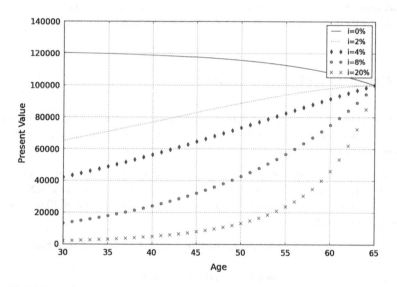

Fig. 4.6 Mathematical reserves as a function of interest rates

have to be calculated separately. The reserve in the case of survival is $V_*(t, A_{**\mathbb{R}})$
and the reserve in the case of death is $V_*(t, A_{*\dagger\mathbb{R}})$ (cf. Definition 4.6.1).

The table of the mathematical reserves indicates that the recursion formula was
used with the boundary condition $x = 65$. Figure 4.6 shows the necessary reserves
for several values of the technical interest rate.

After the calculation of the single premium we will now consider the case of
yearly premiums. The yearly payment of the premiums is modelled by the following
payout function:

Table 4.6 Reserves for an endowment with yearly premiums

Age	Present value premiums	Present value payout	Reserve
65	0	100000	100000
64	−2129	98392	96263
63	−4149	96730	92581
62	−6069	95022	88952
61	−7901	93275	85374
60	−9653	91498	81845
50	−23928	73210	49282
40	−34292	56236	21943
30	−42044	42044	0

$$a_*^{\text{Pre}}(x) = \begin{cases} -P, & \text{if } x < 65, \\ 0, & \text{otherwise.} \end{cases}$$

P has to be calculated such that the value of the insurance is equal to zero at the beginning of the policy. (Equivalence principle: The expected value of the benefits provided by the insurer and the value of the expected premium payments by the insured coincide.) The simplest method to determine the value of P is to consider V_x^{payout} (given by the first two payout functions of the policy) and V_x^{premiums} (given by the third function of the policy, i.e. by the premiums paid in). The total mathematical reserve is then given by $V_x = V_x^{\text{payout}} + V_x^{\text{premiums}}$. But we know that $V_x^{\text{premiums}} = P \times V_x^{\text{premiums}, P=1}$ holds. Thus we can calculate P by the formula

$$P = -V_x^{\text{payout}} / V_x^{\text{premiums}, P=1},$$

since V_x at inception of the policy is 0, as a consequence of the equivalence principle. For our example we get

$$P = 2.129, 15 \text{ USD per year.}$$

Table 4.6 lists the reserves for this insurance with yearly premiums. Figure 4.7 illustrates the same data in a graph.

Exercise 4.8.2 Do the calculation for the previous example.

In the next example we will consider the simple disability insurances model which we looked at earlier. We will show how to model the disability pension with and without an exemption from payment of premiums.

Example 4.8.3 (Disability insurance) We use the model for the disability insurance introduced in Example 2.4.2. Thus in particular we do not incorporate the possibility that insured becomes active again. Moreover we also do not model a waiting period.

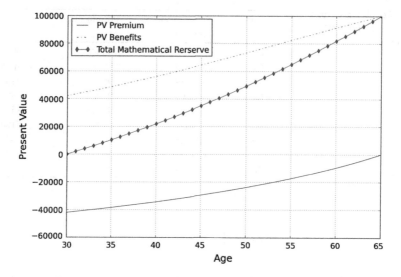

Fig. 4.7 Endowment policy against regular premium payment

- Calculate for a 30-year old man the present value of a (new) disability pension based on 65 as age at maturity and a technical interest rate of 4%.
- Compare for the same person the present value of the premiums for a policy with exemption from payment of premiums and for a policy without this option.

First we calculate the present value of the payouts of the future disability pension. In this case the non-trivial functions which model the policy are: (where we assumed that the disability pension is payable in advance and has the value 1.)

$$a_\diamond^{\text{Pre}} = \begin{cases} 1, & \text{if } x < 65, \\ 0, & \text{otherwise.} \end{cases}$$

The boundary conditions are zero in this case, i.e., there is no payment when the age of maturity is reached. Table 4.7 lists the calculated values for this example. Thus one has to pay 4,396.8 USD as single premium for a future disability pension of 10,000 USD. Furthermore, the reserve is 170,790 USD for a disabled man of 35 years with the same disability pension as above.

Next we consider the present values of the premiums. We have to treat the following two cases separately. On the one hand there is the present value of a policy without exemption from payment of premiums

$$a_*^{\text{Pre}} = \begin{cases} 1, & \text{if } x < 65, \\ 0, & \text{otherwise,} \end{cases}$$

$$a_\diamond^{\text{Pre}} = \begin{cases} 1, & \text{if } x < 65, \\ 0, & \text{otherwise.} \end{cases}$$

Table 4.7 Reserves of
a disability pension

Age	$p_{*\dagger}$	$p_{*\diamond}$	$V_*(x)$	$V_\diamond(x)$
65	0.02289	0.02794	**0.00000**	**0.00000**
64	0.02101	0.02439	0.00000	1.00000
63	0.01929	0.02129	0.02047	1.94299
62	0.01772	0.01860	0.05372	2.83515
61	0.01628	0.01625	0.09427	3.68174
60	0.01495	0.01420	0.13828	4.48719
55	0.00983	0.00732	0.34176	8.01299
50	0.00653	0.00387	0.46175	10.90260
45	0.00439	0.00214	0.50531	13.31967
40	0.00301	0.00127	0.50178	15.35782
35	0.00212	0.00084	0.47493	17.07904
30	0.00155	0.00062	0.43968	18.53012

On the other hand there is the present value of a policy with exemption from payment of premiums ("premium raider")

$$a_*^{\text{Pre}} = \begin{cases} 1, & \text{if } x < 65, \\ 0, & \text{otherwise,} \end{cases}$$

$$a_\diamond^{\text{Pre}} = \begin{cases} 0, & \text{if } x < 65, \\ 0, & \text{otherwise.} \end{cases}$$

The difference between these two policies is, that in the first case also with status "disabled" the insured has to pay the premiums. In both cases the boundary condition at the age of 65 is 0. An application of Thiele's difference equations yields the values listed in Table 4.8. We note that the value of the paid-in premiums is smaller in the case of an exemption of premiums. This can also be seen in the plot of these values in Fig. 4.8.

Now we are able to calculate the yearly premium for a disability insurance of 10,000 USD with exemption from premiums

$$P = 4396.8/18.09044 = 243.05 \text{ USD per year.}$$

Exercise 4.8.4 1. Do the calculations of the above example also for a model which includes the possibility of reactivation (cf. Example 4.2.2).
2. Extend the model by incorporating a waiting period of one year.

Next we consider an insurance on two lives. There are several possible states for which the policy could guarantee a pension.

Example 4.8.5 (Pension on two lives) We start with the calculation of a single premium for several types of an insurance on two lives. For this we assume that the two

Table 4.8 Present value of the premiums

Age	$V(x)$ without premium raider	$V(x)$ with premium raider
65	**0.00000**	**0.00000**
64	1.00000	1.00000
63	1.94299	1.92251
62	2.83515	2.78144
61	3.68174	3.58747
60	4.48719	4.34891
55	8.01299	7.67123
50	10.90260	10.44085
45	13.31967	12.81436
40	15.35782	14.85604
35	17.07904	16.60411
30	18.53012	18.09044

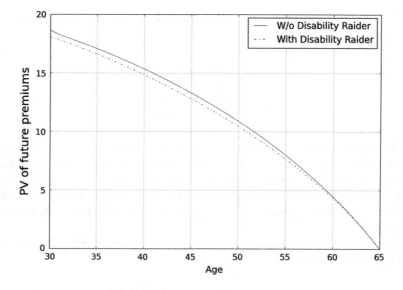

Fig. 4.8 Present value of premiums

persons have the same mortality as given in Example 4.2.1 and that $x_1 = 30$ and $x_2 = 35$ are fixed. We set the technical interest rate to 3.5%, and $\omega = 114$ be the maximal possible age of a living person.

There are three possible pensions: (we denote by x_1 the age of the first person)

Table 4.9 Mathematical reserves (res.) for an insurance on two lives

Notations
$V_{*.}(R1)$ res. for the pension of 1th life, independent of 2nd life,
$V_{.*}(R2)$ res. for the pension of 2nd life, independent of 1th life,
$V_{**}(R1)$ res. for the pension of 1th life, if $X_t = (**)$,
$V_{**}(R2)$ res. for the pension of 2th life, if $X_t = (**)$,
$V_{**}(R\ (**))$ res. for the pension of joint lives, if $X_t = (**)$.

Alter 1	Alter 2	$V_{*.}(R1)$	$V_{.*}(R2)$	$V_{**}(R1)$	$V_{**}(R2)$	$V_{**}(R\ (**))$
115	120	**0.00000**		**0.00000**	**0.00000**	**0.00000**
114	119	1.00000		0.00000	0.00000	1.00000
113	118	1.11505		0.11505	0.00000	1.00000
112	117	1.19991		0.19991	0.00000	1.00000
111	116	1.28640		0.28640	0.00000	1.00000
110	**115**	1.37771	**0.00000**	0.37771	0.00000	1.00000
109	114	1.47450	1.00000	0.45824	0.00000	1.01626
108	113	1.57710	1.11505	0.52973	0.06845	1.04736
90	95	4.64366	3.53796	2.04747	0.94178	2.59618
75	80	9.05696	7.43141	3.39065	1.76510	5.66631
65	70	12.54173	10.77780	3.88770	2.12377	8.65403
55	60	7.78663	6.26733	3.37939	1.86010	4.40724
40	45	4.30964	3.34208	2.13044	1.16288	2.17920
30	35	3.00101	2.30680	1.52353	0.82932	1.47748

State	Type	Formula
$**$	Both persons are alive	$a_{**}^{\mathrm{Pre}}(x) = \alpha_{**} \begin{cases} 0, & \text{if } x_1 < 65, \\ 1, & \text{otherwise,} \end{cases}$
$*\dagger$	Second person is dead	$a_{*\dagger}^{\mathrm{Pre}}(x) = \alpha_{*\dagger} \begin{cases} 0, & \text{if } x_1 < 65, \\ 1, & \text{otherwise,} \end{cases}$
$\dagger*$	First person is dead	$a_{\dagger*}^{\mathrm{Pre}}(x) = \alpha_{\dagger*} \begin{cases} 0, & \text{if } x_1 < 65, \\ 1, & \text{otherwise.} \end{cases}$

The definitions of the pensions above are particular, since the pension for the second life (i.e. for $\dagger*$) is paid at the age of 65. Usually this pension would be paid immediately after the death of the first person.

We set $x = (x_1, x_2)$ and suppose that the two insured die independently. In this case the recursion takes the form

$$V_{**}(x) = a_{**}^{\mathrm{Pre}}(x) + p_{x_1}\, p_{x_2}\, v\, V_{**}(x+1) + p_{x_1}\,(1 - p_{x_2})\, v\, V_{*\dagger}(x+1)$$
$$+(1 - p_{x_1})\, p_{x_2}\, v\, V_{\dagger*}(x+1),$$
$$V_{*\dagger}(x) = a_{*\dagger}^{\mathrm{Pre}}(x) + p_{x_1}\, v\, V_{*\dagger}(x+1),$$
$$V_{\dagger*}(x) = a_{\dagger*}^{\mathrm{Pre}}(x) + p_{x_2}\, v\, V_{\dagger*}(x+1).$$

This recursion yields the values listed in Table 4.9.

Exercise 4.8.6   1. Calculate the present values of the premiums for the insurance
on two lives. Note that also in this calculation one has to consider three different
cases.

2. Create a model for an orphan's pension. For this one has to consider three persons:
father, mother and child. Define the payout functions for a policy which pays
5,000 USD to the child if one of parent dies and 10,000 USD if both die. Assume
that the policy matures if the child is 25 years old.

Chapter 5
Difference Equations and Differential Equations

5.1 Introduction

In this chapter we focus on the Markov model in continuous time. The differential equations are the continuous counter part to the difference equations of the discrete model.

These differential equations where first proved for simple insurance models by Thiele at the end of the 19th century. We are going to derive these equations for the Markov model. They are useful in two ways. On the one hand they help to deepen our understanding of the model. On the other hand they can be used to calculate the premiums for a policy.

5.2 Thiele's Differential Equations

In this section we are going to derive Thiele's differential equations for the mathematical reserve. For simplicity we consider in this chapter only reserves without jumps. Later on we will also allow jumps, but then the proofs become more involved.

Theorem 5.2.1 (Thiele's differential equation) *Let $(X_t)_{t \in T}$, a_{ij}, a_i and δ_t be a regular insurance model (Definition 4.5.6). Moreover, $da_g(t)$ be absolute continuous with respect to the Lebesgue measure λ, i.e. $da_g(t) = a_g(t)\,d\lambda$. (Thus the payout function $A_g(t)$ is continuous.) Then, assuming a deterministic interest intensity, the following statements hold:*

1. *$W_g^+(t)$ is continuous for all $g \in S$.*
2. *$\frac{\partial}{\partial t} W_j^+(t) = -v(t) \left\{ a_j(t) + \sum_{j \neq g \in S} \mu_{jg}(t) a_{jg}(t) \right\}$*
   $+ \mu_j(t) W_j^+(t) - \sum_{j \neq g \in S} \mu_{jg}(t) W_g^+(t)$. (Thiele's differential equation)

M. Koller, *Stochastic Models in Life Insurance*, EAA Series,
DOI: 10.1007/978-3-642-28439-7_5, © Springer-Verlag Berlin Heidelberg 2012

3. $V_j^+(t) = \frac{1}{v(t)} \left[\begin{array}{l} \int_t^u v(\tau)\overline{p}_{jj}(t,\tau)\{a_j(\tau) + \sum\limits_{j \neq g \in S} \mu_{jg}(\tau)(a_{jg}(\tau) \\ + V_g^+(\tau))\}d\tau + v(u)\overline{p}_{jj}(t,u)V_j^+(u^-) \end{array} \right].$

Proof The proof of the first statement is left as an exercise to the reader. To prove the second statement we fix $j \in S$, $t \in \mathbb{R}$ and $\Delta t > 0$. Then Lemma 4.7.2 implies

$$W_j^+(t) = v(t) \left\{ a_j(t) + \sum_{j \neq g \in S} \mu_{jg}(t) a_{jg}(t) \right\} \Delta t$$

$$+ (1 - \mu_j(t)\Delta t) W_j^+(t + \Delta t)$$

$$+ \sum_{j \neq g \in S} \mu_{jg}(t) W_g^+(t + \Delta t) \Delta t + o(\Delta t),$$

where we used the following facts

$$p_{jj}(t, t + \Delta t) = 1 - \Delta t \, \mu_j(t) + o(\Delta t),$$
$$p_{jk}(t, t + \Delta t) = \Delta t \, \mu_{jk}(t) + o(\Delta t).$$

The above equation yields

$$\frac{W_j^+(t + \Delta t) - W_j^+(t)}{\Delta t} = -v(t)\left\{ a_j(t) + \sum_{j \neq g \in S} \mu_{jg}(t) a_{jg}(t) \right\}$$

$$+ \mu_j(t)\, W_j^+(t + \Delta t)$$

$$- \sum_{g \neq j} \mu_{jg}(t) W_g^+(t + \Delta t) + \frac{o(\Delta t)}{\Delta t}.$$

Letting $\Delta t \longrightarrow 0$ we get

$$\frac{\partial}{\partial t} W_j^+(t) = -v(t)\left\{ a_j(t) + \sum_{j \neq g \in S} \mu_{jg}(t) a_{jg}(t) \right\} + \mu_j(t) W_j^+(t)$$

$$- \sum_{j \neq g \in S} \mu_{jg}(t) W_g^+(t).$$

For the proof of the third statement we use Thiele's differential equation:

$$\exp(-\int_o^t \mu_j(\tau)d\tau)\left(-v(t)\{a_j(t) + \sum_{j\neq g \in S} \mu_{jg}(t)a_{jg}(t)\}\right.$$

$$\left. - \sum_{j\neq g \in S} \mu_{jg}(t)W_g^+(t)\right)$$

$$= \exp(-\int_o^t \mu_j(\tau)d\tau)\left(\frac{\partial}{\partial t}W_j^+(t) - \mu_j(t)W_j^+(t)\right)$$

$$= \frac{\partial}{\partial t}\left(\exp(-\int_o^t \mu_j(\tau)d\tau)W_j^+(t)\right).$$

An integration $\int_t^u$ of both sides yields

$$\exp(-\int_o^t \mu_j(\tau)d\tau)\left(\exp(-\int_t^u \mu_j(\tau)d\tau)W_j^+(u) - W_j^+(t)\right)$$

$$= \int_t^u \exp(-\int_o^t \mu_j(\xi)d\xi)\exp(-\int_t^\tau \mu_j(\xi)d\xi)$$

$$\times\left[-v(\tau)\{a_j(\tau) + \sum_{j\neq g \in S} \mu_{jg}(\tau)a_{jg}(\tau)\}\right.$$

$$\left. - \sum_{j\neq g \in S} \mu_{jg}(\tau)W_g^+(\tau)\right]d\tau.$$

Hence, we get

$$V_j^+(t) = \frac{1}{v(t)}\left[\int_t^u v(\tau)\overline{p}_{jj}(t,\tau)\{a_j(\tau) + \sum_{j\neq g \in S} \mu_{jg}(\tau)\right.$$

$$\left. \times (a_{jg}(\tau) + V_g^+(\tau))\}d\tau + v(u)\overline{p}_{jj}(t,u)V_j^+(u^-)\right],$$

where we used that $\overline{p}_{jj}(t,\tau) = \exp(-\int_t^\tau \mu_j(\xi)d\xi)$.

Remark 5.2.2 We derived the following integral equation from Thiele's differential equation:

$$V_j^+(t) = \frac{1}{v(t)}\left[\int_t^u v(\tau)\overline{p}_{jj}(t,\tau)\{\underbrace{a_j(\tau)}_{I} + \sum_{j\neq g \in S} \overbrace{\mu_{jg}(\tau)\underbrace{(a_{jg}(\tau)}_{IIa} +\underbrace{V_g^+(\tau))}_{IIb})}\}d\tau \atop + \underbrace{v(u)\overline{p}_{jj}(t,u)V_j^+(u^-)}_{III}\right].$$

This formula shows the structure of the reserve. The components of the reserve are:

I)   reserve for payments in state j (pensions and premiums),
II)   reserves for state transitions composed of
    IIa)   transition cost (e.g. death benefit) and
    IIb)   necessary reserves in the new state,
III)   reserve, for the case that the insured is still in j after $[t, u]$.

5.3 Examples: Thiele's Differential Equation

In this section we look at examples related to those in the discrete setting. In the first example the differential equations have an explicit solution.

Example 5.3.1 (Term life insurance) We consider a term life insurance with death benefit b, which is financed by a premium of size c. In this situation the differential equations take the following form:

$$\frac{\partial}{\partial t} W_*(t) = v^t (c - \mu_{x+t}\, b) + \mu_{x+t}\, W_*(t) - \mu_{x+t}\, W_{\dagger}(t),$$

$$\frac{\partial}{\partial t} W_{\dagger}(t) = 0$$

with the boundary condition $W_*(s - x) = W_{\dagger}(s - x) = 0$, where s denotes the age of maturity of the policy.

Next, we are going to calculate the mathematical reserve. The above equations obviously imply $W_{\dagger}(t) \equiv 0$. Thus we only have to calculate $W_*(t)$. The homogeneous part of the equation satisfies

$$\frac{dW_*(t)}{W_*(t)} = \mu_{x+t} dt$$

and therefore

$$L_h(t) = A \times \exp\left(\int_0^t \mu_{x+\tau}\, d\tau \right).$$

By variation of constants we get

$$L_p(t) = A(t) \times L_h(t),$$

$$\frac{d}{dt} L_p = A' \times L + A \times L'$$

$$= A' \times L + A \times L$$

$$= A' \times L + L_p,$$

$$A' \times L = v^t (c - \mu_{x+t} \, b),$$

$$A' = v^t (c - \mu_{x+t} \, b) \exp(- \int \mu_{x+\tau} \, d\tau)$$

$$= v^t (c - \mu_{x+t} b) \, _tp_x,$$

$$A(t) = \int_0^t v^\tau (c - \mu_{x+\tau} b) \, _\tau p_x d\tau.$$

Finally, the boundary condition $W_*(s - x) = 0$ yields

$$W_*(s - x) = A(s - x) \times L(s - x)$$

$$= \left[\int_0^{s-x} v^\tau (c - \mu_{x+\tau} b) \, _\tau p_x d\tau \right] \times \left[\exp(\int_0^{s-x} \mu_{x+\tau} d\tau) \right],$$

$$c = b \, \frac{\int_0^{s-x} v^\tau \, _\tau p_x \, \mu_{x+\tau} d\tau}{\int_0^{s-x} v^\tau \, _\tau p_x d\tau}.$$

Example 5.3.2 (Endowment policy) We consider the endowment policy defined in Example 4.8.1. Thus it contains a death benefit of 200,000 USD and an endowment of 100,000 USD. We consider a 30 year old man and 65 as the age of maturity of the policy.

• How much is a single premium for this insurance if the technical interest rate is 3.5%?
• How do these results compare to the values in the corresponding example in discrete time?

We use the mortality rates given by (2.13). For the single premium the following payout function defines the policy:

$$a_{*\dagger}(x) = \begin{cases} 200000, & \text{if } x < 65, \\ 0, & \text{otherwise.} \end{cases}$$

Now Thiele's differential equations are

$$\frac{\partial}{\partial t} W_*(t) = v^t (c - \mu_{x+t} \, a_{*\dagger}(x + t)) + \mu_{x+t} \, W_*(t) - \mu_{x+t} \, W_\dagger(t),$$

$$\frac{\partial}{\partial t} W_\dagger(t) = 0,$$

with the boundary conditions $W_*(s - x) = 100000 \times v(s)$ and $W_\dagger(s - x) = 0$. Then Theorem 5.2.1 yields the results listed in Table 5.1.

Note that the difference of the reserves in the discrete and continuous model have always the same sign. This is caused by the fact, that people only die at the end of the year in the discrete model. Therefore the required single premium is smaller than in the continuous model.

Table 5.1 Discretisation
error for an endowment
policy

Age	$\mu_{*\dagger}(x)$	Reserve discrete model	Reserve cont. model	Diff. in %
65	0.01988	**100000**	**100000**	
64	0.01836	98392	98512	0.12
63	0.01696	96730	96955	0.23
62	0.01566	95022	95341	0.34
61	0.01446	93275	93678	0.43
60	0.01336	91498	91975	0.52
55	0.00897	82360	83092	0.89
50	0.00602	73210	74059	1.16
45	0.00404	64426	65308	1.37
40	0.00271	56236	57096	1.53
35	0.00181	48759	49566	1.65
30	0.00121	42044	42782	1.75

Example 5.3.3   1. Calculate yearly premiums for the previous example.
2. What happens to the discretisation error, if we suppose that people die in the discrete model only at the middle of the year?
3. What happens, if we suppose that the interest rates drops linearly from 6% at the age of 30 to 3% at the age of 65?

Exercise 5.3.4 Calculate with the continuous model the premiums for the policy defined in Example 4.8.3.

Example 5.3.5 (Pensions on two lives) 1. Derive Thiele's differential equation for pensions on two lives.
2. Calculate the premiums for the pensions based on the assumption that the husband and his wife die independently.
3. What happens if the the mortality rate for the state $(**)$ (or $(*\dagger) \cup (\dagger*)$) decreases (or increases) by 15% for each life? (Empirical studies show that the mortality rate of widows and widowers is increased in comparison to the rest of the population.)

For this example we only derive Thiele's differential equations and present the results in a figure. The calculations are done in the setting of Example 4.8.5. (Δt denotes the difference in age of the husband and his wife.)

$$\frac{\partial W_{**}^{+}(t)}{\partial t} = -v^{t}\,a_{**}(t) + (\mu_{x+t}^{husband} + \mu_{x+t+\Delta t}^{wife})\,W_{**}^{+}(t)$$
$$-\mu_{x+t}^{husband}\,W_{\dagger*}^{+}(t) - \mu_{x+t+\Delta t}^{wife}\,W_{*\dagger}^{+}(t),$$
$$\frac{\partial W_{*\dagger}^{+}(t)}{\partial t} = -v^{t}\,a_{*\dagger}(t) + \mu_{x+t}^{wife}\left(W_{*\dagger}^{+}(t) - W_{\dagger\dagger}^{+}(t)\right),$$

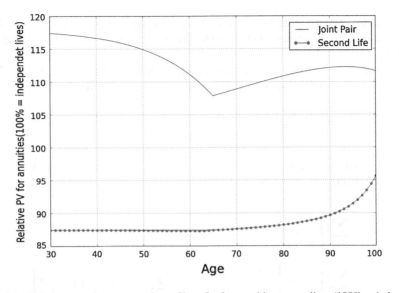

Fig. 5.1 Ratio between the present values of benefits for annuities on two lives (100% = independent mortality probabilities)

$$\frac{\partial W_{\dagger *}^{+}(t)}{\partial t} = -v^{t}\, a_{\dagger *}(t) + \mu_{x+t+\Delta t}^{wife} \left(W_{\dagger *}^{+}(t) - W_{\dagger \dagger}^{+}(t) \right),$$

$$\frac{\partial W_{\dagger \dagger}^{+}(t)}{\partial t} = 0.$$

Figure 5.1 shows the relation of the present values of the benefits for a change in mortality by ±15%. The results are what we expected: a pension on the joint lives becomes more expensive, whereas a pension on the 2nd life becomes cheaper.

Exercise 5.3.6 Complete the previous example.

Exercise 5.3.7 1. Calculate the present value of the premiums for the insurance on two lives. (Also in the continuous setting one has to treat three cases separately.)
2. Create a model for an orphan's pension. In the model one has to consider three persons: farther, mother and child. Define the payout functions for a policy which pays 5,000 USD to the child if one of parent dies and 10,000 USD if both die. Assume that the policy matures if the child is 25 years old.

5.4 Differential Equations for Moments of Higher Order

Thiele's differential equations characterise the mathematical reserve of an insurance policy. In this section we look at the moments of the mathematical reserve. This will enable us for example to calculate the variance of the reserve, which is a measure for its variation. Furthermore, it can be used to analyse the risk structure of an insurance policy.

We start with the difference equations corresponding to the discrete model. The payout functions in the discrete Markov model have the following form:

$$\Delta B_t = \sum_{j\in J} I_j(t)a_j^{Pre}(t) + \sum_{j,k\in J} \Delta N_{jk}(t)\, v_t\, a_{jk}^{Post}(t),$$

where v_t is the yearly discount from $t+1$ to t, with $v_t = \sum_{j\in J} I_j(t)v_j(t)$.
The prospective reserves are

$$V_t^+ = \sum_{\xi=t}^{\infty}(\prod_{k=t}^{k<\xi} v_k)\, \Delta B_\xi$$

$$= \sum_{\xi=t}^{\infty}(\prod_{k=t}^{k<\xi} v_k)\left\{\sum_{j\in J} I_j(\xi)a_j^{Pre}(\xi) + \sum_{j,l\in J} \Delta N_{jl}(\xi)\, v_\xi\, a_{jl}^{Post}(\xi)\right\}.$$

Now our aim is to calculate the expectation of the pth power of the mathematical reserve $((V_t^+)^p)$ conditioned on $\mathcal{F}_t$. The linearity of the integral yields the difference equation

$$V_t^+ = v_t\sum_{j\in J} I_j(t+1)V_{t+1}^+ + \sum_{j\in J} I_j(t)a_j^{Pre}(t) + \sum_{j,k\in J} \Delta N_{jk}(t)\, v_t\, a_{jk}^{Post}(t).$$

This formula indicates that the future reserve is composed of the payments in the period $]t, t+1]$, the payments at time $\{t\}$ and the payments in the period $]t+1, \infty[$. To keep the calculations simple we assume that there is no payment at time $\{t\}$. Furthermore we will use the notation

$$L_t = \sum_{j,k\in J} \Delta N_{jk}(t)\, v_t\, a_{jk}^{Post}(t).$$

Therefore we can simplify the recursion to

$$V_t^+ = v_t\sum_{j\in J} I_j(t+1)V_{t+1}^+ + L_t = \sum_{j\in J} I_j(t+1)\left(v_t\, V_{t+1}^+ + L_t\right).$$

Now the pth moment is given by

$$(V_t^+)^p = \left(\sum_{j \in J} I_j(t+1)(v_t \, V_{t+1}^+ + L_t) \right)^p$$

$$= \sum_{k=0}^{p} \binom{p}{k} \left(\sum_{j \in J} I_j(t+1) v_t \, V_{t+1}^+ \right)^k L_t^{p-k}$$

$$= \sum_{k=0}^{p} \binom{p}{k} \sum_{j \in J} I_j(t+1)(v_t \, V_{t+1}^+)^k L_t^{p-k},$$

where we used the fact that $I_\alpha(t+1) I_\beta(t+1) = \delta_{\alpha\beta} I_\alpha(t+1)$. Next, we can use $P[A \cap B | C] = P[A | B \cap C] \times P[B | C]$ to simplify the recursion for the expectation. We get

$$E\left[(V_t^+)^p \mid X_t = i \right]$$

$$= E\left[\sum_{k=0}^{p} \binom{p}{k} (v_t^i)^k \sum_{j \in J} I_j(t+1)(V_{t+1}^+)^k L_t^{p-k} \mid X_t = i \right]$$

$$= \sum_{k=0}^{p} \binom{p}{k} (v_t^i)^k \sum_{j \in J} E\left[I_j(t+1)(V_{t+1}^+)^k L_t^{p-k} \mid X_t = i \right]$$

$$= \sum_{k=0}^{p} \binom{p}{k} (v_t^i)^k \sum_{j \in J} E\left[I_j(t+1)(V_{t+1}^+)^k \left(v_t \, a_{ij}^{Post}(t) \right)^{p-k} \mid X_t = i \right]$$

$$= (v_t^i)^p \sum_{j \in J} p_{ij}(t, t+1) \sum_{k=0}^{p} \binom{p}{k} \left(a_{ij}^{Post}(t) \right)^{p-k} E\left[(V_{t+1}^+)^k \mid X_{t+1} = j \right]$$

This is the difference equation for the higher order moments of the reserve, if there are no payments in advance. Note that the integration of the payments in advance $a_j^{Pre}(t)$ is not complicated. Nevertheless, to keep the presentation clear we continue without them. We summarise our findings in the following theorem.

Theorem 5.4.1 (Differential equation for moments of higher order) *Under the above assumptions the higher order moments of the reserve satisfy the recursion*

$$E\left[(V_t^+)^p \mid X_t = i \right] = (v_t^i)^p \sum_{j \in J} p_{ij}(t, t+1) \sum_{k=0}^{p} \binom{p}{k} \left(a_{ij}^{Post}(t) \right)^{p-k}$$

$$\times E\left[(V_{t+1}^+)^k \mid X_{t+1} = j \right]$$

Exercise 5.4.2 Derive the above formula for a model which includes payments in advance.

After treating the discrete case we are now going to derive the analogue statements for the continuous setting. The proofs will become more involved, since the model is more general. Before stating the theorem we recall some definitions:

$$dB = \sum_{j \in S} I_j(t)dB_j + \sum_{j \neq k} dB_{jk},$$

$$dv_t = -v_t \times \delta_t \, dt,$$

$$\delta_t = \sum_{j \in S} I_j(t)\delta_j(t).$$

The pth moment of the prospective reserve is defined by

$$V_j^{(p)}(t) := E\left[(V_t^+)^p \mid X_t = j\right]$$

$$= E\left[\left(\frac{1}{v_t}\int_t^\infty v \, dB\right)^p \mid X_t = j\right],$$

where we implicitly assumed that $V_t^+ \in L^p(\Omega, \mathcal{A}, P)$ and that the functions δ, a_i and a_{jk}, μ_{jk} are piecewise continuous. Then the following theorem holds.

Theorem 5.4.3 (Differential equations for moments of higher order) *Under the above assumptions the functions $V_j^{(p)}(t)$ satisfy the differential equation*

$$\frac{\partial}{\partial t}V_j^{(p)}(t) = \left(p\,\delta_j(t) + \sum_{S \ni k \neq j} \mu_{jk}(t)\right)V_j^{(p)}(t) - p\,a_j(t)V_j^{(p-1)}(t)$$

$$- \sum_{j \neq k \in S} \mu_{jk}(t)\sum_{k=0}^p \binom{p}{k}\left(a_{jk}(t)\right)^{p-k} V_j^{(k)}(t)$$

for all $t \in]0, n[\setminus \mathcal{D}$ with the boundary condition

$$V_j^{(p)}(t^-) = \sum_{k=0}^p \binom{p}{k}\left(\Delta a_j(t)\right)^{p-k} V_j^{(k)}(t)$$

for all $t \in \mathcal{D}$. Here $\mathcal{D}$ is the set of discontinuities of the payout function B.

Remark 5.4.4 • The differential equation given above also holds at points where the functions $V_j^{(p)}(t)$ are not differentiable. In these points one gets a valid interpretation by considering the differentials which are given by a formal multiplication with the factor dt.

• The idea of the proof is to represent a suitable martingale in two different ways. These representations will then be used to find a stochastic differential equation for the martingale. Then, since the drift term of the differential equation is zero for a martingale, one obtains an ordinary differential equation.

Proof It turns out to be more convenient to show the differential equation for $W_j^{(p)}(t) = v_t^p \, V_j^{(p)}(t)$. We have

$$dW_j^{(p)}(t) = d(v_t^p) \, V_j^{(p)}(t) + v_t^p \, dV_j^{(p)}(t) \tag{5.1}$$

$$= -p v_t^p \sum_{j \in S} I_j(t) \delta_j(t) dt \, V_j^{(p)}(t) + v_t^p \, dV_j^{(p)}(t). \tag{5.2}$$

Now we define the martingale

$$M^{(p)}(t) := E\left[\left(\int_0^\infty v \, dB \right)^p \bigg| \mathcal{F}_t \right]$$

$$= E\left[\left\{ \left(\int_0^t + \int_t^\infty \right) v \, dB \right\}^p \bigg| \mathcal{F}_t \right]$$

and the function

$$U_t = \int_0^t v \, dB.$$

The Markov property implies

$$E\left[\left\{ \int_t^\infty v \, dB \right\}^{p-k} \bigg| \mathcal{F}_t \right] = \sum_{j \in S} I_j(t) W_j^{(p-k)}(t),$$

and using the Binomial Theorem we get

$$M^{(p)}(t) = \sum_{k=0}^p \binom{p}{k} \sum_{j \in S} U_t^k I_j(t) W_j^{(p-k)}(t).$$

By choosing a right continuous modification of $M^{(p)}$ we can ensure that U and $I_j(t)$ are right continuous.

Now we want to simplify the differential form

$$dM^{(p)}(t) = \sum_{k=0}^p \binom{p}{k} \sum_{j \in S} d\left(U_t^k I_j(t) W_j^{(p-k)}(t) \right).$$

Recall that for a function of bounded variation A we denote by A^{cont} the continuous part and by A^{atom} the discontinuous part. An application of Itô's formula yields

$$d \left(U_t^{(k)} I_j(t) W_j^{(p-k)}(t) \right)$$

$$= k U_t^{(k-1)} dU_t^{cont} I_j(t) W_j^{(p-k)}(t)$$

$$+ U_t^k dI_j^{cont}(t) W_j^{(p-k)}(t) + U_t^k I_j(t) dW_j^{(p-k),cont}(t)$$

$$+ \left\{ U_t^k I_j(t) W_j^{(p-k)}(t) - U_{t^-}^k I_j(t^-) W_j^{(p-k)}(t^-) \right\}. \tag{5.3}$$

To simplify this formula we use for the first line the identities

$$dU_t^{cont} = v_t \sum_{l \in S} I_l(t) a_l(t),$$

$$I_\alpha(t) I_\beta(t) = \delta_{\alpha\beta} I_\alpha(t).$$

For the second line of (5.3) note that the continuous part of $I_\alpha(t)$ vanishes. Finally we have to deal with the jump part in the third line. The jumps have two possible origins. On the one hand they might be caused by a transition:

$$\sum_{j \neq l \in S} \left(\{ U_{t^-} + v_t a_{jl}(t) \}^k W_l^{(p-k)}(t) - U_{t^-}^k W_l^{(p-k)}(t^-) \right) dN_{jl}(t).$$

On the other hand they might be due to a jump in a pension:

$$\sum_{l \in S} I_l(t) \left(\{ U_{t^-} + v_t \Delta a_l(t) \}^k W_l^{(p-k)}(t) - U_{t^-}^k W_l^{(p-k)}(t^-) \right).$$

Moreover, we know that jumps can only occur on the set $\mathcal{D}$ and on this set one can replace $I_l(t)$ by $I_l(t^-)$, since these coincide with probability 1. We also know that a simultaneous jump of both components occurs with probability 0. Finally, $W_l^{(p-k)}(t)$ is continuous and does not induce any jumps, since $\int_t^n v \, dB$ is almost surely continuous on $t \notin \mathcal{D}$. Therefore we get

$$d \left(U_t^{(k)} I_j(t) W_j^{(p-k)}(t) \right)$$

$$= \sum_{l \in S} I_l(t) \left(k U_t^{(k-1)} v_t a_j(t) W_j^{(p-k)}(t) + U_t^k dW_j^{(p-k),cont}(t) \right)$$

$$+ \sum_{j \neq l \in S} \left(\{ U_{t^-} + v_t a_{jl}(t) \}^k W_l^{(p-k)}(t) - U_{t^-}^k W_l^{(p-k)}(t^-) \right) dN_{jl}(t)$$

$$+ \sum_{l \in S} I_l(t^-) \left(\{ U_{t^-} + v_t \Delta a_l(t) \}^k W_l^{(p-k)}(t) - U_{t^-}^k W_l^{(p-k)}(t^-) \right).$$

Applying the fact $X_{t^-} dt = X_t \, dt$ and the previous formula to (5.3) we derive

$$dM^{(p)}(t) - \sum_{j\neq l\in S} \sum_{k=0}^{p} \binom{p}{k} \left(\{U_{t^-} + v_t a_{jl}(t)\}^k W_l^{(p-k)}(t) \right.$$

$$\left. - U_{t^-}^k W_l^{(p-k)}(t^-) \right) dM_{jl}(t)$$

$$= \sum_{j\in S} I_j(t) \sum_{k=0}^{p} \binom{p}{k} \left[k U_t^{(k-1)} v_t a_j(t) W_j^{(p-k)}(t) dt + U_t^k dW_j^{(p-k),cont}(t) \right.$$

$$+ \sum_{j\neq l\in S} \left(\{U_t + v_t a_{jl}(t)\}^k W_l^{(p-k)}(t) - U_t^k W_j^{(p-k)}(t) \right) \mu_{jl}(t) dt \Big]$$

$$+ \sum_{j\in S} I_l(t^-) \sum_{k=0}^{p} \binom{p}{k} \left(\{U_{t^-} + v_t \Delta a_j(t)\}^k W_j^{(p-k)}(t) - U_{t^-}^k W_j^{(p-k)}(t^-) \right),$$

$$(5.4)$$

where we used the identity

$$dM_{ij}(t) = dN_{ij}(t) - I_i(t)\mu_{ij}(t).$$

Note that

$$dW_j^{(p-k),cont}(t) = -(p-k)v_t^{(p-k)}\delta_j(t)\,dt\, V_j^{(p-k)}(t) + v_t^{(p-k)}dV_j^{(p-k),cont}(t).$$

$$(5.5)$$

The left hand side of (5.4) is the differential of a sum of martingales. Thus also the right hand side is the differential of a martingale. Now this has to be constant, since it is previsible and of bounded variation. Therefore the increments of the continuous and the discrete part have to be equal to zero. But this is only possible if

$$0 = \sum_{k=1}^{p} \binom{p}{k} k U_t^{(k-1)} v_t a_j(t) W_j^{(p-k)}(t) dt$$

$$+ \sum_{k=0}^{p} \binom{p}{k} U_t^k W_l^{(p-k)}(t)$$

$$+ \sum_{k=0}^{p} \binom{p}{k} \sum_{j\neq l\in S} \sum_{r=0}^{k} \binom{k}{r} U_t^{(r)} \{v_t a_{jl}(t)\}^{k-r} W_l^{(p-k)}(t)\mu_{jl}(t) dt$$

$$- \sum_{k=0}^{p} \binom{p}{k} U_t^k W_j^{(p-k)}(t) \left(\sum_{j\neq l\in S} \mu_{jl}(t) \right) dt$$

$$(5.6)$$

holds for all $j \in S$ and all $t \in\,]0, n[\setminus \mathcal{D}$. For $x \in \mathcal{D}$ we get

$$0 = \sum_{k=0}^{p} \binom{p}{k} \left(\sum_{r=0}^{k} \binom{k}{r} U_{t^-}^{(r)} \{v_t \Delta a_l(t)\}^{k-r} W_j^{(p-k)}(t) - U_{t^-}^{(k)} W_j^{(p-k)}(t^-) \right).$$

Using the identity

$$\binom{p}{k} k = \binom{p}{k-1} (p - (k-1))$$

we can transform the first line of Eq. (5.6) into

$$\sum_{k=1}^{p} \binom{p}{k-1} (p - (k-1)) U_t^{(k-1)} v_t a_j(t) W_j^{(p-1-(k-1))}(t) dt$$

$$= \sum_{k=0}^{p} \binom{p}{k} (p-k) U_t^{(k)} v_t a_j(t) W_j^{(p-1-k)}(t) dt,$$

where we set $W_j^{(-1)} \equiv 0$. Hence the third line of (5.6) becomes

$$\sum_{k=0}^{p} U_t^{(r)} \sum_{r=0}^{k} \binom{p}{k}\binom{k}{r} \sum_{j \neq l \in S} \{v_t a_{jl}(t)\}^{k-r} W_l^{(p-k)}(t) \mu_{jl}(t) dt.$$

This can be transformed by the identity

$$\binom{p}{k}\binom{k}{r} = \binom{p}{r}\binom{p-r}{k-r}$$

into the representation

$$\sum_{r=0}^{p} \binom{p}{r} U_t^{(r)} \sum_{k=r}^{p} \binom{p-r}{k-r} \sum_{j \neq l \in S} \{v_t a_{jl}(t)\}^r W_l^{(p-k-r)}(t) \mu_{jl}(t) dt,$$

i.e.,

$$\sum_{k=0}^{p} \binom{p}{k} U_t^{(k)} \sum_{r=0}^{p-k} \binom{p-k}{r} \sum_{j \neq l \in S} \{v_t a_{jl}(t)\}^r W_l^{(p-k-r)}(t) \mu_{jl}(t) dt.$$

If we now gather the powers of U_t we get

$$0 = \sum_{k=0}^{p} \binom{p}{k} U_t^{(k)} dQ_j^{(p-k)}(t),$$

where

$$dQ_j^{(q)}(t) = qv_t a_j(t) W_j^{(q-1)}(t)dt + dW_j^{(q),cont}(t)$$

$$+ \sum_{k=0}^{q} \binom{q}{k} \sum_{j\neq l\in S} \{v_t a_{jl}(t)\}^k \, W_l^{(q-k)}(t)\mu_{jl}(t)dt$$

$$- W_j^{(q)}(t) \sum_{j\neq l\in S} \mu_{jl}(t)\,dt.$$

This equation implies $dQ_j^{(0)}(t) \equiv 0$ and thus $dQ_j^{(q)}(t) \equiv 0$ by induction. Finally, the formula above and Equations (5.1) and (5.5) imply the result.

5.5 The Distribution of the Mathematical Reserve

The distribution function can be used to answer questions which only depend on the tail of the distribution. Thus it is important for the estimation of extreme risks.

This section has the same structure as the previous section. In the beginning, we solve the problem for the discrete time set. Afterwards we treat the continuous time model.

Recall that the cash flows in the discrete Markov model are given by

$$\Delta B_t = \sum_{j\in J} I_j(t)a_j^{Pre}(t) + \sum_{j,k\in J} \Delta N_{jk}(t) v_t a_{jk}^{Post}(t).$$

We want to calculate the distribution function of the discounted future cash flows

$$P_i(t, u) = P\left[\sum_{j=t}^{\infty} (\prod_{k=t}^{k<j} v_k)\, \Delta B_j < u | X_t = i \right].$$

The following identities hold:

$$P_i(t, u) = P\left[\sum_{j=t}^{\infty} D_{t,j}\, \Delta B_j < u | X_t = i \right]$$

$$= \sum_{l\in J} p_{il}(t)\, P\left[\sum_{j=t}^{\infty} D_{t,j}\, \Delta B_j < u | X_t = i, X_{t+1} = l \right]$$

$$= \sum_{l \in J} p_{il}(t) P\left[v_{i,t} \sum_{j=t+1}^{\infty} D_{t,j} \, \Delta B_j < u - a_i^{Pre}(t) \right.$$

$$\left. - v_{i,t} \, a_{il}^{Post}(t) | X_{t+1} = l \right]$$

$$= \sum_{l \in J} p_{il}(t) \, P_l(t+1, v_{i,t}^{-1}(u - a_i^{Pre}(t)) - a_{il}^{Post}(t)), \text{ where}$$

$$D_{t,j} = \prod_{k=t}^{k<j} v_k.$$

These relations are summarised in the following theorem.

Theorem 5.5.1 (Distribution of the reserves) *The distribution function of the reserves satisfy the recursion*

$$P_i(t, u) = \sum_{j \in J} p_{ij}(t) \, P_j(t+1, v_{i,t}^{-1}(u - a_i^{Pre}(t)) - a_{ij}^{Post}(t)).$$

Besides the recursion formula also boundary conditions are required. These are, in contrast to the previous problems, now given in form of distributions rather than fixed values. For an insurance whose mathematical reserve is equal to zero at maturity the boundary condition is for example given by: (Here ω denotes the maximal age at which insured persons are alive.)

$$P_i(\omega + 1, u) = \begin{cases} 0, & \text{if } u \leq 0, \\ 1, & \text{if } u > 0. \end{cases}$$

The distribution function of a pension in the discrete model is shown in Fig. 5.2. The jumps, which are caused by the discrete model, are clearly visible.

In the previous sections we have seen that one can use differential equations to find the moments of the reserves. For cash flows of sufficient regularity one could also prove that the moments are differentiable.

Analogous one could try to find differential equations for the distribution functions. But the following example shows, that for distribution functions this is not an easy task.

Example 5.5.2 This example will illustrate that the distribution function can be discontinuous even for relatively simple insurance policies. We consider an endowment policy, with a death benefit of 100,000 USD and an endowment of 200,000 USD. Now we want to calculate the reserve for an insured of 30 years of age. Here T_x will denote the future life span. The following equations hold:

$$V_{30} = \begin{cases} 100000 \times v^{T_x}, & \text{if } T_x < 35, \\ 200000 \times v^{65-30}, & \text{if } T_x \geq 35. \end{cases}$$

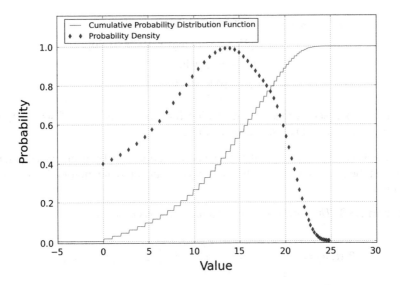

Fig. 5.2 Probability distribution function for the present value of an immediate payout annuity $(x = 65)$

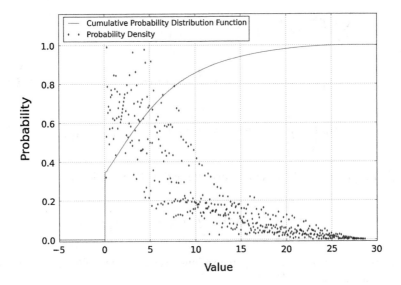

Fig. 5.3 Probability distribution function of the present value of a deferred widows pension $(x = 65)$

Now assume a technical interest rate of 1.5%. Then the following equation holds for $0 < \alpha \leq 100000$:

$$P[V_{30} < \alpha] = P[100000v^{T_x} < \alpha, \ T_x < 35]$$
$$= P[35 > T_X > \log(\alpha/100000)/\log(v)]$$
$$= {}_\gamma p_x - {}_{35} p_x, \text{ where}$$
$$\gamma = \log(\alpha/100000)/\log(v).$$

This calculation shows that the distribution function has a jump of size ${}_{35}p_x$ at $200000v^{35}$, and thus it is discontinuous (Fig. 5.3).

The next theorem shows that the distribution functions satisfy an integral equation. Note that the recursion in discrete time provides an approximation to this integral equation.

Theorem 5.5.3 *The conditional distribution functions of the reserves*

$$P_j(t, u) = P\left[\int_t^\infty \exp\left(-\int_t^\xi \delta_\tau d\tau \right) dB(\xi) \le u \,|\, X_t = j \right]$$

satisfy the integral equation

$$P_j(t, u) = \sum_{k \ne j} \int_t^\infty \exp\left(-\int_t^\xi \sum_{l \ne j} \mu_{jl}(\tau)d\tau \right) \mu_{jk}(\xi)$$

$$\times P_k\left(\xi, \exp(\delta_j(\xi - t))u - \int_t^\xi \exp(\delta_j(\xi - \tau))dB_j(\tau) - a_{jk}(s) \right) d\xi$$

$$+ \exp\left(-\int_t^\infty \sum_{l \ne j} \mu_{jl}(\tau)d\tau \right) \chi_{[\int_t^n \exp(-\delta_j(\tau-t))dB_j(\tau) \le u]}. \qquad (5.7)$$

Proof The proof is analogous to the discrete setting. One considers

$$A = \left\{ \int_t^\infty \exp\left(-\int_t^\xi \delta \right) dB(\xi) \le u \right\}$$

and treats, as in the discrete setting, the various cases separately. We leave the proof to the reader and refer to [HN96].

Exercise 5.5.4 Complete the proof of the previous theorem [HN96].

After deriving the integral equations for the distribution function, we will modify these slightly. The equations are still hard to handle, since the right hand side depends on t. To overcome this problem we define

$$Q_j(t, u) := P_j\left(t, \exp(\delta_j t)\left(u - \int_0^t \exp(-\delta_j \tau)dB_j(\tau) \right) \right).$$

The mapping from P to Q can be inverted by

$$P_j(t, u) = Q_j\left(t, \exp(-\delta_j t)u + \int_0^t \exp(-\delta_j \tau)dB_j(\tau)\right).$$

Using Q_j one can easily derive the equation

$$\exp\left(-\int_0^t \mu_j\right) Q_j(t, u) = \int_t^n \exp\left(-\int_0^s \mu_j\right) \sum_{k \neq j} \mu_{jk}(s)$$

$$\times Q_k\left(s, \exp((\delta_j - \delta_k)s)u + \int_0^s \exp(-\delta_k \tau)dB_k(\tau)\right.$$

$$\left. - \exp((\delta_j - \delta_k)s) \int_0^s \exp(-\delta_j \tau)dB_j(\tau) - \exp(-\delta_k s)a_{jk}(s)\right)ds$$

$$+ \exp\left(-\int_0^n \mu_j\right) \chi_{\left[\int_0^n \exp(-\delta_j(\tau))dB_j(\tau) \leq u\right]}.$$

Theorem 5.5.5 *The functions Q satisfy (in the sense of Stieltje's differentials) the following differential equations*

$$d_t Q_j(t, u) = \mu_j \, dt \, Q_j(t, u) - \sum_{k \neq j} \mu_{jk} \, dt$$

$$\times Q_k\left(t, \exp((\delta_j - \delta_k)t) u + \int_0^t \exp(-\delta_k \tau)dB_k(\tau)\right.$$

$$\left. - \exp((\delta_j - \delta_k)t) \int_0^t \exp(-\delta_j \tau)dB_j(\tau) - \exp(-\delta_k t) a_{jk}(t)\right),$$

with the boundary conditions

$$Q_j(n, u) = \chi_{\left[\int_0^n \exp(-\delta_j \tau)dB_j(\tau) \leq u\right]}.$$

Remark 5.5.6 Theorem 5.5.5 proves useful, since the equations given therein are easier to solve by numerical methods. The main idea is to derive Q in a first step and then calculate P based on Q.

Chapter 6
Examples and Problems From Applications

6.1 Introduction

In this chapter we take a closer look at problems which appear in applications. Moreover, the examples will show the scope of the Markov model and illustrate some special modelling tricks. The examples are based on the discrete model, since this is most popular in applications.

Besides the structural conditions given by the model one has to consider the actual conditions fixed in the policies. This is necessary since on the one hand the model should also be able to represent existing policies, on the other hand formulas are often based on explicit contract terms. Therefore, in the following we will try to understand the formulas based on the policy setup.

6.2 Monthly and Quarterly Payments

We start with the problem of monthly and quarterly payments. For this note that in applications one often considers payment periods of one year, but the actual payments are done monthly or quarterly. As an example we look at a 4/4 pension with payments in advance. That is a pension which pays $\frac{1}{4}$ quarterly. Let us suppose for the moment that the mortality rate for a given year is q_x and that the quarterly mortality $q_x^{[4]}$ is given by

$$(1 - q_x^{[4]})^4 = 1 - q_x.$$

Thus the mortality is constant during the whole year. The current, paid in advance, pension for the discrete time model with time intervals of 3 month is given by

$$a_*(t) = \frac{1}{4},$$

M. Koller, *Stochastic Models in Life Insurance*, EAA Series,
DOI: 10.1007/978-3-642-28439-7_6, © Springer-Verlag Berlin Heidelberg 2012

and we have the recursion

$$V_*(t) = a_*(t) + (1 - q_x^{[4]})v^{\frac{1}{4}} V_*\left(t + \frac{1}{4}\right).$$

Now we can derive the recursion for a time interval of one year. We get

$$V_*(t) = a_*(t) + (1 - q_x^{[4]})v^{\frac{1}{4}} V_*\left(t + \frac{1}{4}\right)$$

$$= a_*(t) \times \sum_{k=0}^{3} \left((1 - q_x^{[4]})v^{\frac{1}{4}}\right)^k + (1 - q_x)v V_*(t+1)$$

$$= a_*(t) \times \frac{1 - (1 - q_x)v}{1 - \left((1 - q_x^{[4]})v^{\frac{1}{4}}\right)} + (1 - q_x)v V_*(t+1)$$

$$\approx \frac{5}{8} + (1 - q_x)v \left(\frac{3}{8} + V_*(t+1)\right),$$

where, in the last step, the following Taylor expansion of f at $z = 1$ was used:

$$f(z) = \frac{1}{4}\left(1 + z^{0.25} + z^{0.50} + z^{0.75}\right),$$

$$f(1) = 1,$$

$$\frac{d}{dz} f(z)|_{z=1} = \frac{3}{8},$$

$$f(z) \approx 1 + \frac{3}{8}(z - 1)$$

$$= \frac{5}{8} + \frac{3}{8}z.$$

The above example shows that one models quarterly payments also by a model with a time interval of one year. Note that the approximation which we derived is in fact the same as the one which is used for a model with commutation functions. Moreover one should note, that these arguments can also be applied to other policies, e.g. insurances on two lives.

Exercise 6.2.1 1. How do you adapt the method from above, which uses a time interval of one year, to a disability pension with a 3 month waiting period?
2. Calculate the corresponding approximation for a quarterly pension on two lives (for the state **) with payments in advance.
3. One can also solve the previous example by decomposing the exact solution into two terms of the form $(1 - q_x)$ and v, respectively. What is the solution in this case?

Table 6.1 Approximation error for a quarterly pension

x	Part per year exact	Part per year approx.	Total exact	Total approx.	Error total (%)
114	0.4436	0.6421	0.4436	0.6421	44.7533
113	0.5298	0.6681	0.5808	0.7420	27.7533
112	0.5874	0.6922	0.6915	0.8253	19.3363
111	0.6323	0.7145	0.7973	0.9115	14.3198
110	0.6692	0.7351	0.9033	1.0027	11.0058
105	0.7917	0.8170	1.4893	1.5472	3.8884
100	0.8614	0.8724	2.2214	2.2597	1.7228
95	0.9047	0.9098	3.1358	3.1630	0.8654
90	0.9325	0.9351	4.2484	4.2687	0.4773
85	0.9508	0.9521	5.5575	5.5733	0.2846
80	0.9629	0.9636	7.0436	7.0564	0.1817
75	0.9709	0.9714	8.6713	8.6820	0.1231
70	0.9763	0.9766	10.3937	10.4028	0.0879
65	0.9799	0.9801	12.1588	12.1667	0.0656
60	0.9823	0.9825	13.9156	13.9227	0.0508
50	0.9850	0.9851	17.2341	17.2399	0.0335

Note, the formula above does not require that all payments within the year are of the same amount.

Example 6.2.2 Now we calculate the size of the error in the mathematical reserve, which is caused by the approximation of a quarterly pension with payments in advance. We use the mortalities defined by (2.13). The results are listed in Table 6.1. This shows that in the usual range, up to 85 years, the error is very small.

6.3 Pensions with Guaranteed Payment Periods

Now we want to consider the problem of pensions with a guaranteed minimal number of payment periods. This type of insurance is offered, since an insured does not want that all his paid in premiums are lost in the case of an early death. This means that, when reaching the age of maturity of the policy, the insured is guaranteed a fixed number of pension payments. Technically this corresponds to an adaptation of the mortality rate for the insured during the guarantee period.

Another approach to this problem is the following. We consider a pension with guaranteed payment periods (10 years guaranteed after the age of 65). Hence 65 is the age of maturity of the policy. For the standard pension the non trivial functions which define the policy are given by

$$a_*(t) = \begin{cases} 0, & \text{if } t < 65, \\ 1, & \text{if } t \geq 65. \end{cases}$$

For the pension with guarantee one has to split the state † into dying before 65 (denoted by: $\dagger_<$) and dying after 65 ($\dagger_\ge$). In this case the relevant transition probabilities are

$$p_{**}(x) = 1 - q_x,$$

$$p_{*\dagger_<}(x) = \begin{cases} q_x, & \text{if } t < 65, \\ 0, & \text{if } t \ge 65, \end{cases}$$

$$p_{*\dagger_\ge}(x) = \begin{cases} 0, & \text{if } t < 65, \\ q_x, & \text{if } t \ge 65, \end{cases}$$

$$p_{\dagger_<\dagger_<}(x) = 1,$$

$$p_{\dagger_\ge\dagger_\ge}(x) = 1.$$

The non trivial functions which define the policy take now the following form (For a 1/1 pension with payments in advance.)

$$a_*(t) = \begin{cases} 0, & \text{if } t < 65, \\ 1, & \text{if } t \ge 65, \end{cases}$$

$$a_{\dagger_\ge}(t) = \begin{cases} 1, & \text{if } t \in [65; 75[, \\ 0, & \text{otherwise.} \end{cases}$$

We will give an example which illustrate these pensions with a guaranteed minimal number of payments.

Example 6.3.1 This example shows the development of the mathematical reserve for a 65 year old man. We assume a guarantee period of 15 years. (Mortality rate given by GRM 1995 (see Appendix D), 1/1 pension with payments in advance, $\omega = 121$).

Age	Premium with guarantee	Premium without guarantee	Ratio in (%)
121	10000	10000	100
120	14694	14694	100
110	27143	27143	100
100	42031	42031	100
90	65298	65298	100
80	91840	91840	100
75	124057	107387	116
70	151184	124817	121
65	174024	142454	122

Figure 6.1 illustrates the mathematical reserve of a term pension (20 years) with 10 years guarantee period, which starts to payout immediately.

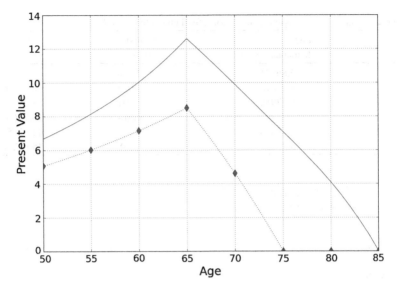

Fig. 6.1 Guaranteed annuity ($\diamond$ = guaranteed part)

6.4 Refund Guarantee

A pension with refund guarantee is a special policy, which guaranties a refund of a certain part of the premiums to the insured in case of an early death. There are several possibilities for the guaranteed refund:

1. refund of the paid in premiums only before maturity of the policy,
2. refund of the difference of paid in premiums and paid out benefits, (total refund)
3. refund of the current mathematical reserve before or after the beginning of the pension payments.

One can understand the first two types of refund guarantees as an additional death benefit. These types of refund guarantees have been part of policies for a long time, and they are well studied. Therefore we will concentrate here on the third type of refund guarantee. Also this type of refund can be modelled in several ways.

Example 6.4.1 (Mathematical reserve refund guarantee) We want to calculate the premiums for the policy with a mathematical reserve refund guarantee. But first we consider an insurance whose death benefit coincides with the mathematical reserve. In this case the following recursion holds:

$$V_*(x) = 1 + p_{**}(x)\, v\, V_*(x+1) + p_{*\dagger}(x)\, V_*(x).$$

(We assumed that the mathematical reserve for the refund guarantee is paid at the beginning of the period.) Now a simple transformation yields the modified recursion equation

Table 6.2 Comparison of single premiums for refund guarantees (KT 1995, $s = 65$, male)

Age	Refund mathematical reserve	Refund premium	Total refund
40	5.86921	5.50680	5.58673
45	6.97078	6.64053	6.78605
50	8.27910	8.01191	8.27932
55	9.83297	9.65914	10.15667
60	11.67848	11.61117	12.55151
65	13.87038	13.87038	15.69276

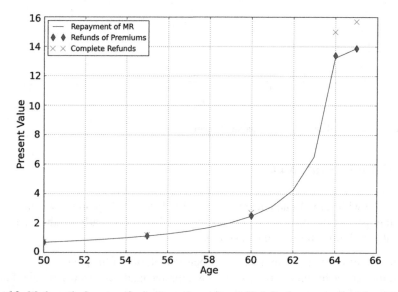

Fig. 6.2 Mathematical reserve for permanent pensions with refund guarantee based on yearly premiums

$$V_*(x) = \frac{1 + p_{**}(x)\, v\, V_*(x + 1)}{(1 - p_{*\dagger}(x))}.$$

This solves the problem in the case of a single premium. To calculate yearly premiums one has to solve a more complicated equation, which can be done easily by numerical methods.

Table 6.2 shows a comparison of single premiums for pensions with various types of refund guarantee. In the case of refund of the mathematical reserve the guarantee ends when the pension payments start. Figure 6.2 illustrates the same comparison for yearly premiums.

Example 6.4.2 This example will be a bit more involved than the previous. We consider a widow's pension with a single premium. Let additionally the policy guarantee

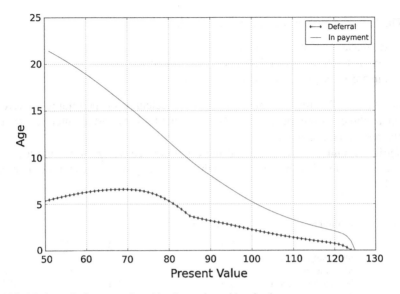

Fig. 6.3 Mathematical reserve of a widow's pension with refund guarantee

a refund of the mathematical reserve until the husband is 85 years, which is paid if the wife dies or in case of a simultaneous (in the same year) death of the couple. The corresponding recursion is

$$V_{(**)}(x) = \frac{v\,(p_{(**)(**)}\,V_{(**)}(x+1) + p_{(**)(\dagger*)}\,V_{(\dagger*)}(x+1))}{(1 - p_{(**)(*\dagger)}(x) - p_{(**)(\dagger\dagger)}(x))}.$$

Figure 6.3 depicts the solution to this equation. It clearly indicates that the refund guarantee only contributes until 85. Afterwards the shown solution is just based on the standard recursion equation.

6.5 Insurances with Stochastic Interest Rate

In this section we consider insurances with stochastic interest rate. We want to focus on the application of methods which were introduced above. Only exemplary interest rate models are used. They are not meant to resemble reality. We consider the endowment policy defined in Example 4.2.1. Thus we have a policy for a 30 year old man, with a death benefit of 200,000 USD and an endowment of 100,000 USD. The aim is to calculate both: the single premium and yearly premiums.

The following interest rate models are used:

1. a technical interest rate fixed at 5%,
2. an interest rate which models a periodic economy,
3. an interest rate given by a random walk model.

Example 6.5.1 (Interest rate models) In order to calculate the premiums we have to set up the three interest rate models. This is trivial for the constant interest rate, and for the remaining two note:

Periodic economy: We assume that the economy behaves periodic and a period consists of 8 states. The interest rate and the transition probabilities for the states are given by

State	Comment	Interest rate (%)	p_{ii}	p_{ii+1}	p_{ii+2}
0	Start	5.0	0.1	0.7	0.2
1	Increasing interest	5.5	0.1	0.7	0.2
2	Max. interest	6.0	0.1	0.7	0.2
3	Decreasing interest	5.5	0.1	0.7	0.2
4	Average	5.0	0.1	0.7	0.2
5	Decreasing interest	4.5	0.1	0.7	0.2
6	Min. interest	4.0	0.1	0.7	0.2
7	Increasing interest	4.5	0.1	0.7	0.2

This model yields a periodic interest rate, where the duration of a period is random.
Random walk: We consider a random walk on a finite set, whose transition probabilities are modified at the boundary:

State	Comment	Interest rate (%)	p_{ii-1}	p_{ii}	p_{ii+1}
0	Min. interest	4.0	0.0	0.5	0.5
1		4.3	0.4	0.2	0.4
2		4.7	0.4	0.2	0.4
3	Starting position	5.0	0.4	0.2	0.4
4		5.3	0.4	0.2	0.4
5		5.7	0.4	0.2	0.4
6	Max. interest	6.0	0.5	0.5	0.0

In both models we assume that the current interest rate is 5%.

We are going to simplify the model by using the stochastic interest rate only for the transition $* \rightsquigarrow *$. Thus we always use the technical interest rate of 5% for the transition $* \rightsquigarrow \dagger$.

Example 6.5.2 (Single and yearly premiums) To calculate the premiums we apply the recursions to the interest rate models. This yields:
Fixed interest rate: In this case the premium is $P = 24755/16.77946 = 1475.30$ USD per year, and we get the following results (PV = present value):

Age	Benefits PV	Premiums PV	Mathematical reserve Single premium	Mathematical reserve Yearly premiums
65	100000	0.00000	100000	100000
64	96510	1.00000	96510	95035
60	83599	4.45585	83599	77026
55	69535	7.84428	69535	57963
50	57483	10.49434	57483	42000
45	47219	12.59982	47219	28631
40	38554	14.28589	38554	17478
35	31305	15.64032	31305	8231
31	25974	16.57022	25974	1528
30	24755	16.77946	24755	0

Periodic interest rate: In this case the premium is $P = 24630/16.65234 = 1479.07$ USD per year, and we get:

Age	Benefits PV $i_t = 4\%$	Benefits PV $i_t = 5\%$	Benefits PV $i_t = 6\%$	Mathematical reserve Yearly premiums $i_t = 5\%$
65	100000	100000	100000	100000
64	97414	95624	96510	95035
60	83854	83369	82536	76004
55	70204	68904	68923	57431
50	57834	57170	57128	41766
45	47482	46996	46812	28368
40	38832	38316	38233	17323
35	31517	31130	31072	8181
31	26145	25838	25761	1509
30	24914	24630	24558	0

Random walk: In this case the premium is $P = 24936/16.81204 = 1483.20$ USD per year, and we get:

Age	Benefits PV $i_t = 4\%$	Benefits PV $i_t = 5\%$	Benefits PV $i_t = 6\%$	Mathematical reserve Yearly premiums $i_t = 5\%$
65	100000	100000	100000	100000
64	97414	96510	95624	95027
60	86558	83611	80775	77002
55	73429	69588	65924	57951
50	61470	57581	53886	42008
45	50934	47356	43963	28652
40	41854	38717	35746	17503
35	34154	31482	28952	8247
31	28476	26155	23959	1531
30	27171	24936	22822	0

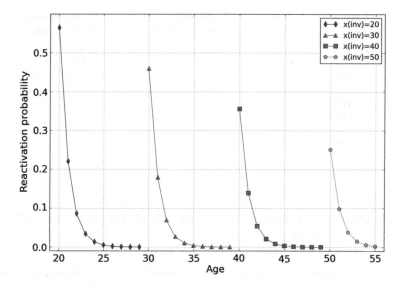

Fig. 6.4 Reactivation probability

6.6 Disability Insurance

We will look at a term disability insurance based on a Markov model. The model for a disability insurance consists at least of three states $\{*, \diamond, \dagger\}$ and the corresponding transition probabilities.

Since the reactivation probability (i.e. the probability of becoming active after a disability incurred) depends on the duration of the disability, we going to decompose the state $\diamond$ into states $\diamond_1, \diamond_2, \ldots, \diamond_n$. Here the state $\diamond_k$ represents persons whose duration of disability is in $[k-1, k[$. The state $\diamond_n$ has a special meaning: we suppose that persons in this state have a permanent disability. The temporal behaviour of the reactivation probabilities is shown in Fig. 6.4. Note that the initial reactivation probability is high for a young person, but it decreases with the age of the insured.

Moreover for any fixed age at which the disability occurs, the reactivation probability decreases more or less exponentially. Because of this rapid drop in the reactivation probability, one often agrees on some initial waiting period. These induce a minor modification to the model. The state space with the possible transitions for the disability insurance model is show in Fig. 6.5.

Since we want to decompose $\diamond$ into the states $\diamond_1, \ldots, \diamond_n$ we are faced with an obvious question: how large must be n such that the error of the model is less than a given bound? We will answer this question for a model with the following transition probabilities:

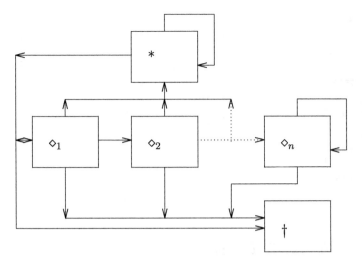

Fig. 6.5 State space diagram for a disability insurance

$$p_{*\dagger}(x) = \exp(-7.85785 + 0.01538x + 0.000577355x^2),$$
$$p_{*\diamond_1}(x) = 3 \times 10^{-4} \times (8.4764 - 1.0985x + 0.055x^2),$$
$$p_{\diamond_k *}(x) = \begin{cases} \exp(-0.94(k-1)) \times \alpha(x,k), & \text{if } k < n, \\ 0, & \text{otherwise,} \end{cases}$$
$$\alpha(x,k) = 0.773763 - 0.01045(x - k + 1),$$
$$p_{\diamond_k \dagger}(x) = 0.008 + p_{*\dagger}(x),$$
$$p_{**}(x) = 1 - p_{*\diamond_1}(x) - p_{*\dagger}(x),$$
$$p_{\diamond_k \diamond_{k+1}}(x) = 1 - p_{\diamond_k *}(x) - p_{\diamond_k \dagger}(x).$$

For the calculations we further assume that we consider a 1/1 disability pension with payments in advance and with policy function

$$a_{\diamond_k}^{\text{Pre}}(x) = \begin{cases} 1, & \text{if } x < 65, \\ 0, & \text{otherwise.} \end{cases}$$

The mathematical reserve for an active person for different values of n is show in Fig. 6.6.

It turns out that for $n = 6$ the error for people of 25 to 65 years of age is less than 5%. Figure 6.7 shows the mathematical reserve for different ages for the disability insurance model with $n = 6$.

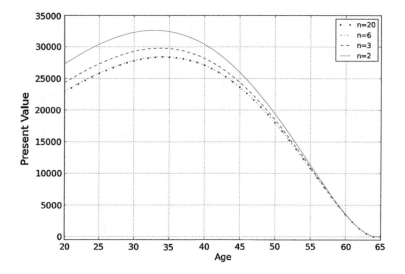

Fig. 6.6 Mathematical reserve for active persons for various n

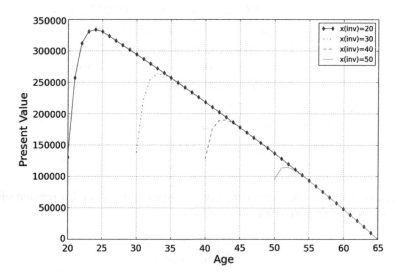

Fig. 6.7 Mathematical reserve for disabled persons for $n = 6$

Chapter 7
Hattendorff's Theorem

7.1 Introduction

Hattendorff's Theorem (1868) states that the losses which incur for an insurance policy in different years are uncorrelated and have mean zero. When this theorem was first discovered it caused many discussions. Nowadays it is part of every introduction to stochastic modelling in insurance. In the following we present the theorem in its general form and its version for the Markov model.

7.2 Hattendorff's Theorem: General Setting

We start with a cash flow B which is discounted by v. Thus the present value is given by $V = \int v_t \, dB$. Further we assume that v and B are adapted to the filtration $\mathbb{F} = (\mathcal{F}_t)_{t \geq 0}$. Then the prospective mathematical reserve $V_{\mathbb{F}}^+(t)$ is given by:

$$V_{\mathbb{F}}^+(t) = E\left[\frac{1}{v(t)} \int_{]t,\infty]} v \, dB \,\Big|\, \mathcal{F}_t\right].$$

Definition 7.2.1 (*Loss for the insurer*) The *loss for the insurer* in the time interval $]s,t]$ (with $s < t$) discounted to the time 0 of a cash flow B is defined by

$$L_{]s,t]} := \int_{]s,t]} v \, dB + v(t) \, V_{\mathbb{F}}^+(t) - v(s) \, V_{\mathbb{F}}^+(s).$$

The loss is composed of the following parts:

$\int_{]s,t]} v \, dB$	payments in the interval $]s,t]$,
$v(t) \, V_{\mathbb{F}}^+(t)$	value of the policy at the end of the period,
$-v(s) \, V_{\mathbb{F}}^+(s)$	value of the policy at the beginning of the period.

M. Koller, *Stochastic Models in Life Insurance*, EAA Series,
DOI: 10.1007/978-3-642-28439-7_7, © Springer-Verlag Berlin Heidelberg 2012

Remark 7.2.2 The main tool in the proof of Hattendorff's Theorem is the fact that

$$M(t) = E\left[V\middle|\mathcal{F}_t\right]$$

$$= \int_{[0,t]} v \, dB + v(t) \, V_{\mathbb{F}}^+(t)$$

is an $\mathbb{F}$-adapted martingale, which we can assume (without loss of generality) to be right continuous. Hence, L can be represented by M in the following way:

$$L_{]s,t]} = M(t) - M(s).$$

Definition 7.2.3 Let $(T_i)_{i\in\mathbb{N}}$ be a non decreasing sequence of stopping times $(0 = T_0 < T_1 < \ldots)$. Then we set

$$L_i^T = L_{]T_{i-1}, T_i]}.$$

Remark 7.2.4 The most important sequences of stopping times $(T_i)_{i\in\mathbb{N}}$ are:

1. $T_i = \alpha_i$ is an increasing sequence of numbers (e.g. the ends of the policy periods),
2. the T_i are the times of a state transition.

To unify the notation, we also set

$$L_0 = B(0) + V_{\mathbb{F}}(0) - E[V].$$

Theorem 7.2.5 (Hattendorff) *Let* $V \in L^2(dP)$, *i.e.* V *has finite moments up to order 2. Then the following statements are equivalent:*

1. $E\left[L_i^T\right] = 0$ *and* $(L_i^T)_{i\in\mathbb{N}}$ *are uncorrelated,*
2. $E\left[L_j^T\middle|\mathcal{F}_i\right] = 0$ *and* $Cov\left(L_j^T, L_k^T\middle|\mathcal{F}_i\right) = 0$ *for all* $i < j, k$,
3. $Var\left(\sum_{k=j}^{\infty} L_k^T\middle|\mathcal{F}_{T_i}\right) = \sum_{k=j}^{\infty} Var\left(L_k^T\middle|\mathcal{F}_{T_i}\right)$,
4.

$$Var\left(L_j^T\middle|\mathcal{F}_{T_i}\right) = E\left[[M, M](T_j) - [M, M](T_{j-1})\middle|\mathcal{F}_{T_i}\right]$$

$$= E\left[\int_{]T_{j-1}, T_j]} d[M, M]\middle|\mathcal{F}_{T_i}\right],$$

where $d[M, M](t) = E\left[(dM(t))^2\middle|\mathcal{F}_{t-}\right]$.

The first two statements are due to Hattendorff.

Proof $M(t)$ is a martingale, since $V \in L^1(dP)$. Thus $L_{]s,t]} = M(t) - M(s)$ holds, since B and v are $\mathbb{F}$-adapted. Now we can apply Doob's optional sampling theorem to the martingale M, i.e. for stopping times $S \leq T$ we get

$$M_S = E\left[M_T \Big| \mathcal{F}_S\right].$$

1. Let the stopping times $S \leq T$ be given. Then

$$
\begin{aligned}
E[L_{]S,T]}] &= E[M_T - M_S] \\
&= E\left[E\left[M_T - M_S \Big| \mathcal{F}_S\right]\right] \\
&= E[M_S - M_S] = 0.
\end{aligned}
$$

   For the second part of the statement we consider the corresponding stopping times $S \leq T \leq U \leq V$ and find

$$
\begin{aligned}
E&\left[L_{]S,T]}L_{]U,V]}\right] \\
&= E\left[(M_T - M_S)(M_V - M_U)\right] \\
&= E\left[M_T E\left[M_V - M_U \Big| \mathcal{F}_T\right]\right] - E\left[M_S E\left[M_V - M_U \Big| \mathcal{F}_S\right]\right] \\
&= E[M_T(M_T - M_T)] - E[M_S(M_S - M_S)] = 0.
\end{aligned}
$$

2. This is essentially proved in the same way as the the first statement.
3. For a finite sum the statement is a consequence of the previous two. Furthermore, $X \mapsto E[X|\mathcal{F}]$ is continuous. Thus the statement is a consequence of the monotone convergence theorem.
4. Let $U \leq S \leq T$ be stopping times such that $L_i = M_T - M_S$. We define

$$
H_t = \begin{cases} 1, & \text{if } t \in]S, T], \\ 0, & \text{otherwise} \end{cases}
$$
$$
= 1_{[0,T]} - 1_{[0,S]}.
$$

   Hence $L_i = (H \cdot M)_\infty$ holds and we get

$$
\begin{aligned}
\text{Var}\left(L_j \Big| \mathcal{F}_{T_i}\right) &= E\left[(H \cdot M)_\infty^2 \Big| \mathcal{F}_{T_i}\right] \\
&= E\left[[M, M](T_j) - [M, M](T_{j-1}) \Big| \mathcal{F}_{T_i}\right] \\
&= E\left[\int_{]T_{j-1}, T_j]} d[M, M] \Big| \mathcal{F}_{T_i}\right].
\end{aligned}
$$

Exercise 7.2.6 Complete the proof of the theorem above.

7.3 Hattendorff's Theorem: Markov Model

Now we want to state Hattendorff's Theorem for the Markov model which we developed in the previous chapters. We concentrate on a regular insurance model (Definition 4.5.6) for which the payout functions a_i are absolute continuous with respect to the Lebesgue measure and the interest intensities ($\delta_i = \delta$) are deterministic.

Definition 7.3.1 Let $j \in S$ and $n \in \mathbb{N}$. Then $S_n^{(j)}$ denotes the nth arrival time in state j. Similarly, $T_n^{(j)}$ denotes the nth departure time from state j.

S_n and T_n are stopping times, since S is discrete. Based on these stopping times we can define the total loss in state j:

Definition 7.3.2 For each $j \in S$ we set

$$L^j = \sum_{n=1}^{\infty} \left[M(T_n^{(j)}) - M(S_n^{(j)}) \right]$$

$$= \sum_{n=1}^{\infty} \int_{]S_n^{(j)}, T_n^{(j)}]} dM.$$

The following lemma is needed in order to determine the variance of the loss.

Lemma 7.3.3   1. $\{X(t^-) = j\} = \bigcup_{n=1}^{\infty} \{S_n^{(j)} < t \le T_n^{(j)}\}$.
2. $L^j = \int \chi_{\{X(t^-)=j\}} dM$.

3.
$$Var\left(L^j\right) = E\left[\int_0^{\infty} \chi_{\{X(t^-)=j\}} (dM(t))^2 \right]$$

$$= \int_0^{\infty} E\left[\chi_{\{M(t^-)=j\}} E\left[dM[M, M] \Big| \mathcal{F}_{t^-} \right] \right].$$

Proof   1.   The proof is left as an exercise to the reader.
2.   This is a consequence of the relation $\int_A dM = \int \chi_A dM$.
3. $L^j = \int_0^{\infty} \chi_{\{X(t^-)=j\}} dM$ holds and therefore

$$Var(L^j) = E\left[\left[L^j, L^j \right]_{\infty} \right] \text{ by [Pro90] Cor. 2.6.4}$$

$$= E\left[\int_0^{\infty} \chi_{\{X(t^-)=j\}}^2 d[M, M] \right] \text{ by [Pro90] Thm. 2.6.28}$$

$$= E\left[\int_0^{\infty} \chi_{\{X(t^-)=j\}} d[M, M] \right]$$

$$= \int_0^{\infty} E\left[\chi_{\{X(t^-)=j\}} E\left[(dM(t))^2 | \mathcal{F}_{t^-} \right] \right],$$

where the last equality holds in general, under certain assumptions. See also [Nor96a, Nor92].

We are going to apply the previous lemma to the Markov model, but first recall the following statements:

Remark 7.3.4 • $[A + B, A + B] = [A, A] + 2[A, B] + [B, B]$,
• $|[A, B]| \leq \sqrt{[A, A]} \times \sqrt{[B, B]}$,
• $[A, B] = AB - \int A_- dB - \int B_- dA$.
• A is of bounded variation $\Longleftrightarrow [A, A] = 0$.
• For a process of quadratic variation $\Delta[A, A] = (\Delta A)^2$ holds.

Next, we are going to calculate $d[M, M]$ and $d[M, M]|\mathcal{F}_{t^-}$. We know that

$$M(t) = \int_0^t v(\tau) \left[\sum_{j \in S} a_j(\tau) I_j(\tau) d\tau + \sum_{S \ni k \neq j} a_{jk}(\tau) dN_{jk}(\tau) \right]$$

$$+ E \left[\sum_{j \in S} I_j(t) W_j(t) \Big| \mathcal{F}_{t^-} \right], \quad \text{where}$$

$$W_j(t) = v(t) V_j(t)$$

and

$$M(t) = A_0(0) + \int_{]0,t]} \left(\sum_j I_j(\tau) dA_j(\tau) + \sum_{j \neq k} a_{jk}(\tau) dN_{jk}(\tau) \right)$$

$$+ \sum_j I_j(t) W_j(t). \tag{7.1}$$

To proceed we will use partial integration for semimartingales. For this consider the variation process, and note that

$$[I_j, W_j] = 0,$$

since on the one hand $[W_j, W_j] = 0$ (due to $t \mapsto W_j(t)$ being of bounded variation) and on the other hand

$$0 \leq |[I_j, W_j]| \leq \sqrt{[I_j, I_j]} \times \sqrt{[W_j, W_j]} = 0.$$

Thus partial integration yields

$$I_j(t) W_j(t) = I_j(0) W_j(0) + \int_{]0,t]} I_j(\tau) dW_j(\tau) + \int_{]0,t]} W_j(\tau^-) dI_j(\tau).$$

To calculate dI_j we can use the identities

$$I_j(t) = \sum_{k \neq j} \left(N_{kj}(t) - N_{jk}(t) \right)$$

and

$$dI_j(t) = \sum_{k \neq j} \left(dN_{kj}(t) - dN_{jk}(t) \right).$$

Finally, we get

$$\sum_j I_j(t) W_j(t) = \sum_j I_j(0) W_j(0) + \sum_j \int_{]0,t]} I_j(\tau) dW_j(\tau)$$

$$+ \sum_j \int_{]0,t]} W_j(\tau^-) \sum_{j \neq k} \left(dN_{kj}(\tau) - dN_{jk}(\tau) \right)$$

$$= W_0(0) + \sum_j \int_{]0,t]} I_j(\tau) dW_j(\tau)$$

$$+ \sum_{j \neq k} \int_{]0,t]} \left(W_k(\tau^-) - W_j(\tau^-) \right) dN_{jk}(\tau).$$

This equation tells us, that the change of the prospective mathematical reserve is equal to

- the change of the mathematical reserve, if there is no state transition
- the difference of the old and the new mathematical reserve corresponding to a transition.

Now, by combining the formula which we derived and Eq. (7.1) we get the following theorem.

Theorem 7.3.5 *The martingale M can be calculated by the formula*

$$M(t) = A_0(0) + W_0(0) + \int_0^t \sum_j I_j(t) d \left(A_j(\tau) + W_j(\tau) \right)$$

$$+ \sum_j \int_{]0,t]} \sum_{k \neq j} \left(a_{jk}(\tau) + W_k(\tau^-) - W_j(\tau^-) \right) dN_{jk}(\tau).$$

Based on this we can calculate the quadratic variation of M. First of all we know that A and W are continuous and we introduce the notations:

$$A := \int_0^t \sum_j I_j(t) d\left(A_j(\tau) + W_j(\tau)\right),$$

$$B := \sum_j \int_{]0,t]} \sum_{k \neq j} \left(a_{jk}(\tau) + W_k(\tau^-) - W_j(\tau^-)\right) dN_{jk}(\tau).$$

Then

$$[M, M] = [A + B, A + B] = [A, A] + 2[A, B] + [B, B],$$

where $|[A, B]| \leq \sqrt{[A, A]} \times \sqrt{[B, B]}$. Next, we define $Z_j = A_j(\tau) + W_j(\tau)$ and note that Z is continuous, since

$$\begin{aligned}
Z_j(\tau^-) &= A_j(\tau^-) + W_j(\tau^-) \\
&= A_j(\tau) - \left(A_j(\tau) - A_j(\tau^-)\right) + W_j(\tau^-) \\
&= A_j(\tau) + W_j(\tau) \\
&= Z_j(\tau).
\end{aligned}$$

The continuity of $Z_j(\tau)$ implies $[Z_j, Z_j] = 0$ and therefore also the following statements hold:

$$[A, B] = 0,$$
$$[A, A] = 0.$$

These equations yield the following theorem.

Theorem 7.3.6 (Hattendorff)

$$d[M, M](t) = \sum_{k \neq j} \left(a_{jk}(t) + V_k(t^-) - V_j(t^-)\right)^2 v(t)^2 dN_{jk}(t).$$

Proof The statement is simple if we recall that we defined $v(\tau) = 1$. Then we just have to rewrite the previous equations with the substitutions

$$a_{jk} \longrightarrow v(\tau) a_{jk}(\tau),$$
$$v(\tau) V_k(\tau) = W_k(\tau).$$

The value

$$R_{jk}(t) = a_{jk}(t) + V_k(t^-) - V_j(t^-)$$

is called value at risk. It is composed of

1. the payment, which is due when a transition from j to k occurs,
2. the difference of the mathematical reserves corresponding to the transition.

A further consequence of the calculations above is the next theorem, which provides a formula for the variance of the loss.

Theorem 7.3.7 *The following statements hold:*

1. Let $X(0) = 1$, then

$$Var\, L^j(0) = \int_{]0,\infty[} v(\tau)^2 p_{1j}(0,\tau) \sum_{j \neq k} \mu_{jk}(\tau) R_{jk}^2(\tau) d\tau.$$

2. The variance of the future loss $L_j^k(t)$, conditioned on $X(t) = j$, is

$$Var\, L_j^k(t) = \frac{1}{v(t)^2} \int_{]t,\infty[} v(\tau)^2 \, p_{jk}(t,\tau) \sum_{l \neq k} \mu_{kl}(\tau) R_{kl}^2(\tau) d\tau.$$

Chapter 8
Unit-Linked Policies

8.1 Introduction

Up to now we have mostly considered models with deterministic interest rate or with an interest rate given by a Markov chain on a finite state space. This helped us to keep the calculations simple. In this chapter we have a look at some more general models. On the one hand we consider models for policies whose actual value depends on the performance of an underlying unit (usually a fond), on the other hand we will discuss further models with stochastic interest rate.

With respect to these two types of models one has to note, that the corresponding theory is still in development and no final presentation is available. Especially, up to now, there is no mutual consent on the models used for the interest rate processes. In fact there are various opinions on the ideal model for the stochastic interest rate, and each of these models yields a different numerical result.

What are the advantages and disadvantages of these models? As mentioned before, models try to resemble reality and they are more or less precise. Thus clearly the following relation holds: if the model has more stochastic elements, then

- it might resemble reality more precisely,
- it also becomes more complex with respect to assumptions, parameter fitting and numerical algorithms.

In our opinion, the main difficulty when using a stochastic interest rate is due to the diversity of ideas for the possible models. In the following we do not want to rate these models, but our aim is to describe a general approach which provides a fundamental knowledge of this area and of the required techniques.

We begin with a look at policies whose value is tied to a bond or a fond. The payout of theses so called "unit-linked policies" usually consists of a certain number of shares of a fond (the underlying unit) to the insured, in case of an occurrence of the insured event.

These policies have the characteristic feature that the benefits (endowments or death benefits) are not deterministic, but random. A unit-linked policy is usu-

M. Koller, *Stochastic Models in Life Insurance*, EAA Series,
DOI: 10.1007/978-3-642-28439-7_8, © Springer-Verlag Berlin Heidelberg 2012

ally financed by a single premium. This type of financing is preferred due to the management of these policies. Note that this is in contrast to traditional policies. Moreover one has to note, that the value at risk is constant for a traditional policy, but for a unit-linked policy it depends on the underlying fond.

To analyse unit-linked policies, we introduce the following notation:

$$N(t) \quad \text{number of shares at time } t,$$
$$S(t) \quad \text{value of a share at time } t.$$

We assume that $N(t)$ is deterministic. The relevant quantities for a life insurance in this setting and in the traditional setting are summarised in the following table:

	traditional	(pure) unit-linked
death benefit value (time 0)	$C(t) = 1$ $\pi_0(t) = \exp(-\delta t)$	$C(t) = S(t)$ $\pi_0(t) = S(0)$ (assuming a "normal" economy)
single premium	$E\left[\int_0^T \pi_0(t) d(\chi_{T_x \le t})\right]$ $= \int_0^T \exp(-\delta t)_t p_x \mu_{x+t} dt$	$E\left[\int_0^T \pi_0(t) d(\chi_{T_x \le t})\right]$ $= S(0) \int_0^T {}_t p_x \mu_{x+t} dt$ $= (1 - {}_T p_x) S(0)$

The above terms indicate that the financial risk taken by the insurer is smaller for a unit-linked product than for a traditional product with fixed technical interest rate.[1] Furthermore, note that in the calculation of the single premium we implicitly assumed that the mean of the discounted (to time 0) value of the fond at time t coincides with the value of the fond at time 0. This means, that first of all we have to start with a discussion of the value or price of a fond.

The model also did not include any guarantees. But in general one would like to add a guarantee (e.g. a refund guarantee for the paid in premiums) to the policy. For example, the guarantee could be of the form

$$G(t) = \int_0^t \bar{p}(s) ds,$$

where $\bar{p}(s)$ denotes the density of the premiums at time s. More general one could be interested in a refund guarantee of the paid in premiums with an additional interest at a fixed rate:

$$G(t) = \int_0^t \exp(r(t - s)) \bar{p}(s) ds.$$

---

[1] Actually, this is not true in general. Here we implicitly assumed that the capital market risks of a unit-linked policy are minimised by an appropriate trading strategy. For a classical insurance such a trading strategy replicates the cash flows by zero coupon bonds with the corresponding maturities.

In these examples the payout function would be

$$C(t) = \max(S(t), G(t)).$$

Let us assume that the value of the fond is given by a stochastic process (with distribution P). What is, in this setting, the value of the discounted payment $C(t)$ at time t? A first guess might be

$$\pi_0\left(C(t)\right) = E^P\left[\max(S(t), G(t))\right].$$

But it is not that simple! If this would be the value of the payment, there would be the possibility to make a profit without risk (arbitrage). In order to prevent this possibility one has to change the measure P. In mathematical finance it is proved that an equivalent martingale measure exists, such that there is no arbitrage. Then in a "fair" market we have

$$\pi_0\left(C(t)\right) = E^Q\left[\max(S(t), G(t))\right],$$

where Q is a measure equivalent to P such that the discounted value of the underlying fond is a martingale.

Furthermore, in mathematical finance payments like $C(t)$ are called the payouts of an option. To determine the price of an option, one uses the arbitrage free pricing theory. A quick introduction to this theory will be given in the next section.

8.2 Pricing Theory

In this section we have a look at modern financial mathematics. It is not our aim to give a comprehensive exposition with proofs of every detail, which would easily fill a whole book. We only want to give a brief survey which illustrates the theory. The reader interested in more details is referred to [Pli97, HK79, HP81, Duf92].

In this context we clearly also have to mention the paper of Black and Scholes [BS73] with their famous formula for pricing of options.

8.2.1 Definitions

First of all we start with an example which illustrates the use of the pricing theory. The price of a share, modelled by a geometric Brownian motion $(S_t(\omega))$, might develop as shown in Fig. 8.1.

A European call option for a certain share is the right to buy these shares at a fixed price c (strike price) at a fixed time T. The value of this right at time T is

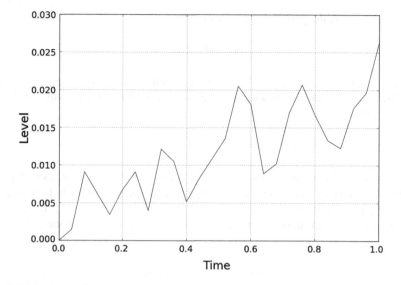

Fig. 8.1 Movement of a share price

$$H = \max\left(S_T - c, 0\right).$$

Now a bank would like to know the value (i.e., the fair price) of this option at time 0. As noted in the previous section, taking the expectation would systematically yield the wrong values. In many cases this would provide the possibility to make profits without risk. Thus there would be arbitrage opportunities.

To simplify the exposition we will consider the most simple models of the economy, i.e. finite models. In particular also the time set will be discrete. The reader interested in the corresponding theorems for continuous time can for example find these in [HP81]. In the following the ideas and concepts of the pricing theory are presented.

Let $(\Omega, \mathcal{A}, P)$ be a probability space where Ω is a finite set. Moreover, we assume that $P(\omega) > 0$ holds for all $\omega \in \Omega$.

We also fix a finite time T, as the time at which all trading is finished. The σ-algebra of the observable events at time t is denoted by $\mathcal{F}_t$, and the shares are traded at the times $\{0, 1, 2, \ldots, T\}$.

We suppose that there are $k < \infty$ stochastic processes, which represent the prices of the shares $1, \ldots, k$, i.e.,

$$S = \{S_t, t = 0, 1, 2, \ldots, T\} \text{ with components } S^0, S^1, \ldots S^k.$$

As usual, we assume that each S^j is adapted to $(\mathcal{F}_t)_t$. Here S_t^j can be understood as the price of the jth share at time t. The fact that the price process has to be adapted reflects the necessity that one has to know at time t the previous price of S. The share

S^0 plays a special role. We suppose that $S_t^0 = (1 + r)^t$, i.e., we have the possibility to make risk free investments which provide interest rate r. The risk free discount factor is defined by

$$\beta_t = \frac{1}{S_t^0}.$$

Next, we are going to define what is meant by a trading strategy.

Definition 8.2.1 A *trading strategy* is a previsible ($\phi_t \in \mathcal{F}_{t-1}$) process $\Phi = \{\phi_t, t = 1, 2, \ldots, T\}$ with components ϕ_t^k.

We understand ϕ_t^k as the number of shares of type k which we own during the time interval $[t - 1, t[$. Therefore ϕ_t is called the portfolio at time $t - 1$.

Notation 8.2.2 *Let X, Y be vector valued stochastic processes. Then we use the notations:*

$$< X_s, Y_t > = X_s \cdot Y_t = \sum_{k=0}^{n} X_s^k \times Y_t^k,$$

$$\Delta X_t = X_t - X_{t-1}.$$

Next, we want to determine the value of the portfolio at time t:

time	value of the portfolio
$t - 1$	$\phi_t \cdot S_{t-1}$
t^-	$\phi_t \cdot S_t$

Thus, the return in the interval $[t - 1, t[$ is $\phi_t \cdot \Delta S_t$, and hence the total return is

$$G_t(\phi) = \sum_{\tau=1}^{t} \phi_\tau \cdot \Delta S_\tau.$$

We fix $G_0(\phi) = 0$, and $(G_t)_{t \geq 0}$ is called return process.

Theorem 8.2.3 *G is an adapted and real valued stochastic process.*

Proof The proof is left as an exercise to the reader.

Definition 8.2.4 A trading strategy is *self financing*, if

$$\phi_t \cdot S_t = \phi_{t+1} \cdot S_t, \quad \forall t = 1, 2, \ldots, T - 1.$$

A self financing trading strategy is just a trading strategy where at no time further money is added to or deduced from the portfolio.

Definition 8.2.5 A trading strategy is *admissible*, if it is self financing and

$$V_t(\phi) := \begin{cases} \phi_t \cdot S_t, & \text{if } t = 1, 2, \ldots, T, \\ \phi_1 \cdot S_0, & \text{if } t = 0 \end{cases}$$

is non negative. (In other words, one is not allowed to become bankrupt.) The set of admissible trading strategies is denoted by Φ.

Remark 8.2.6 The idea of admissible trading strategies is to consider only portfolios which neither lead to bankruptcy nor allow an addition or deduction of money. This also indicates, that the value of the trading strategies remains constant when the portfolio is rearranged. Thus a trading strategy, which generates the same cash flow as an option, can be used to determine the value of the option.

Definition 8.2.7 A contingent claim is a positive random variable X. The set of all contingent claims is denoted by $\mathcal{X}$.

A random variable X is attainable, if there exists an admissible trading strategy $\phi \in \Phi$ which replicates it, i.e.

$$V_T(\phi) = X.$$

In this case one says "ϕ replicates X".

Definition 8.2.8 The price of an attainable contingent claim, which is replicated by ϕ, is denoted by

$$\pi = V_0(\phi)$$

(We will see later, that this price is not necessarily unique. It coincides with the initial value of the portfolio.)

8.2.2 Arbitrage

We say, the model offers arbitrage opportunities, if there exists

$$\phi \in \Phi \text{ with } V_0(\phi) = 0 \text{ and } V_T(\phi) \text{ positive and } P[V_T(\phi) > 0] > 0,$$

i.e., money is generated out of nothing. If such a strategy exists, one can make a profit without taking any risks. One of the axioms of modern economy says, that there are no arbitrage opportunities. This is fundamental for some important facts in the option pricing theory.

Now, we are going to define what is meant by a price system.

Definition 8.2.9 A mapping

$$\pi : \quad \mathcal{X} \to [0, \infty[, \quad X \mapsto \pi(X)$$

is called *price system* if and only if the following conditions hold:

- $\pi(X) = 0 \iff X = 0$,
- π is linear.

A price system is *consistent*, if

$$\pi(V_T(\phi)) = V_0(\phi) \quad \text{for all } \phi \in \Phi.$$

The set of all consistent price systems is denoted by Π, and $\mathbb{P}$ denotes the set

$$\mathbb{P} = \{Q \text{ is a measure equivalent to } P, \text{ s.th. } \beta \times S \text{ is a martingale w.r.t. } Q\},$$

where β is the discount factor from time t to 0. The measures $\mu \in \mathbb{P}$ are called equivalent martingale measures.

Theorem 8.2.10 *There is a bijection between the consistent price systems $\pi \in \Pi$ and the measures $Q \in \mathbb{P}$. It is given by*

1. $\pi(X) = E^Q[\beta_T X]$.
2. $Q(A) = \pi(S_T^0 \chi_A)$ *for all* $A \in \mathcal{A}$.

Proof Let $Q \in \mathbb{P}$. We define $\pi(X) = E^Q[\beta_T X]$. Then π is a price system, since P is strictly positive on Ω and Q is equivalent to P. Thus it remains to show, that π is consistent. For $\phi \in \Phi$ we get

$$\beta_T V_T(\phi) = \beta_T \phi_T S_T + \sum_{i=1}^{T-1} (\phi_i - \phi_{i+1}) \beta_i S_i$$

$$= \beta_1 \phi_1 S_1 + \sum_{i=2}^{T} \phi_i (\beta_i S_i - \beta_{i-1} S_{i-1}),$$

where we used that ϕ is self financing. This yields

$$\pi(V_T(\phi)) = E^Q[\beta_T V_T(\phi)]$$

$$= E^Q[\beta_1 \phi_1 S_1] + E^Q\left[\sum_{i=2}^{T} \phi_i (\beta_i S_i - \beta_{i-1} S_{i-1})\right]$$

$$= E^Q[\beta_1 \phi_1 S_1] + \sum_{i=2}^{T} E^Q\left[\phi_i E^Q\left[(\beta_i S_i - \beta_{i-1} S_{i-1}) | \mathcal{F}_{i-1}\right]\right]$$

$$= \phi_1 E^Q[\beta_1 S_1]$$

$$= \phi_1 \beta_0 S_0,$$

since ϕ is previsible and βS is a martingale with respect to Q.

Thus, π is a consistent price system.

Now let $\pi \in \Pi$ be a consistent price system and Q be defined as above. Then $Q(\omega) = \pi(S_t^0 \chi_{\{\omega\}}) > 0$ holds for all $\omega \in \Omega$, since $S_t^0 \chi_{\{\omega\}} \neq 0$. Moreover, we have $\pi(X) = 0 \iff X = 0$ and therefore Q is absolutely continuous with respect to P.

In the next step, we are going to show that Q is a probability measure. We define

$$\phi^0 = 1 \quad \text{and} \quad \phi^k = 0 \; \forall k \neq 0.$$

Hence, by the consistency of π, we get

$$\begin{aligned}
1 &= V_0(\phi) \\
&= \pi(V_T(\phi)) \\
&= \pi(S_T^0 \cdot 1) \\
&= Q(\Omega).
\end{aligned}$$

The prices of positive contingent claims are positive and Q is additive. Therefore, Kolmogorov's axioms are satisfied, since Ω is finite. We have $Q(\omega) = \pi(S_T^0 \cdot \chi_{\{\omega\}})$ by definition. Hence, also

$$E[f] = \sum_\omega \pi(S_T^0 \cdot \chi_{\{\omega\}}) \cdot f(\omega) = \pi\left(S_T^0 \cdot \sum_\omega f(\omega)\right).$$

Thus, with $f = \beta_t X$, we have

$$E^Q[\beta_T X] = \pi(S_T^0 \cdot \beta_T \cdot X) = \pi(X).$$

Now we still have to show that $\beta_T S_T^k$ is a martingale for all k. Let k be a coordinate and τ be a stopping time, and set

$$\begin{aligned}
\phi_t^k &= \chi_{\{t \leq \tau\}}, \\
\phi_t^0 &= \left(S_\tau^k / S_\tau^0\right) \chi_{\{t > \tau\}}.
\end{aligned}$$

(We keep the share k up to time τ, then it is sold and the money is used for a risk free investment.) It is easy to show, that the strategy ϕ is previsible and self financing. Finally, for an arbitrary stopping time τ,

$$\begin{aligned}
V_0(\phi) &= S_0^k, \\
V_T(\phi) &= \left(S_\tau^k / S_\tau^0\right) S_T^0
\end{aligned}$$

and

$$\begin{aligned}
S_0^k &= \pi(S_T^0 \cdot \beta_T \cdot S_\tau^k) \\
&= E^Q\left[\beta_\tau \cdot S_\tau^k\right].
\end{aligned}$$

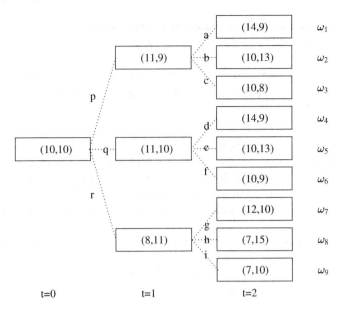

Fig. 8.2 Example calculation of an option price

Thus $\beta_T\, S_T^k$ is a martingale with respect to Q.

Above we have proved one of the main theorems in the option pricing theory. Next, we will present further statements without proofs. They all can be found for example in [HP81].

Theorem 8.2.11 *The following statements are equivalent*

1. *There is no arbitrage opportunity,*
2. $\mathbb{P} \neq \emptyset$,
3. $\Pi \neq \emptyset$.

Lemma 8.2.12 *Suppose there exists a self financing strategy $\phi \in \Phi$ such that*

$$V_0(\phi) = 0, \ V_T(\phi) \geq 0, \ E[V_T(\phi)] > 0.$$

Then there exists an arbitrage opportunity.

Example 8.2.13 We are going to calculate the price of an option for a simple example. Consider a market with two shares $Z = (Z_1, Z_2)$ which are traded at the times $t = 0$, $t = 1$ and $t = 2$. Figure 8.2 shows the possible behaviour of these shares in form of a tree. To calculate the price of the option we suppose that all nine possibilities have the same probability.

We want to calculate the price of a complex option given by

$$X = \{2Z_1(2) + Z_2(2) - [14 + 2\min(\min\{Z_1(t), Z_2(t)\}, 0 \leq t \leq 2)]\}^+.$$

First of all, we have to find an equivalent martingale measure. Thus we have to solve
for the times $t = 0$ and $t = 1$ the following equations:

$$10 = 11p + 11q + 8r, \quad \text{(martingale condition for } Z_1)$$
$$10 = 9p + 10q + 11r, \quad \text{(martingale condition for } Z_2)$$
$$1 = p + q + r.$$

The solution to these equations is $p = q = r = \frac{1}{3}$.

Here we can see explicitly which circumstances imply the existence and uniqueness of a martingale measure. In this example the martingale measure is, from a
geometric point of view, defined as the intersection of three hyper-planes. Depending on their orientation, there is either one or there are many or there is none equivalent
martingale measure.

Next, we can derive the equations for the times $t = 1$ and $t = 2$. These are

$$11 = 14a + 10b + 10c,$$
$$9 = 9a + 13b + 8c,$$
$$1 = a + b + c,$$
$$11 = 14d + 10e + 10f,$$
$$10 = 9d + 13e + 9f,$$
$$1 = d + e + f,$$
$$8 = 12g + 7h + 7i,$$
$$11 = 10g + 15h + 10i,$$
$$1 = g + h + i,$$

and they are solved by $(a, b, c) = (0.25, 0.15, 0.60)$, (d, e, f)
$= (0.25, 0.25, 0.50)$ and $(g, h, i) = (0.20, 0.20, 0.60)$.

Now we know the transition probabilities with respect to the martingale measure,
which enables us to calculate the martingale measure Q itself. The results of these
calculations are summarised in the following table:

state	$X(\omega_i)$	$Q(\omega_i)$
ω_1	5	1/12
ω_2	1	1/20
ω_3	0	1/5
ω_4	5	1/12
ω_5	0	1/12
ω_6	0	1/6
ω_7	4	1/15
ω_8	1	1/15
ω_9	0	1/5

Finally, we can calculate the price of the option as expectation with respect to Q.
The result is $\frac{73}{60}$.

8.2.3 Continuous Time Models

For models in continuous time we restrict our exposition to the statements, the proofs can be found in the references mentioned before. A major difference between the discrete and the continuous setting is that we are going to *assume* that $\mathbb{P} \neq \emptyset$ holds for the continuous time model.

We start with some basic definitions.

Definition 8.2.14 • A trading strategy ϕ is a locally bounded, previsible process.
• The value process corresponding to a trading strategy ϕ is defined by

$$V : \Pi \to \mathbb{R}, \phi \mapsto V(\phi) = \phi_t \cdot S_t = \sum_{i=0}^{k} \phi_t^k \cdot S_t^k.$$

• The return process G is defined by

$$G : \Pi \to \mathbb{R}, \phi \mapsto G(\phi) = \int_0^\tau \phi \, dS = \int_0^\tau \sum_{i=0}^{k} \phi^k dS^k.$$

• ϕ is self financing, if $V_t(\phi) = V_0(\phi) + G_t(\phi)$.
• To define admissible trading strategies we use the notation:

$$Z_t^i = \beta_t \cdot S_t^i, \qquad \text{discounted value of share } i$$

$$G^*(\phi) = \int \sum_{i=1}^{k} \phi^i \, dZ^i, \qquad \text{discounted return}$$

$$V^*(\phi) = \beta \, V(\phi) = \phi^0 + \sum_{i=1}^{k} \phi^i \, Z^i.$$

A trading strategy is called admissible, if it has the following three properties:

1. $V^*(\phi) \geq 0$,
2. $V^*(\phi) = V^*(\phi)_0 + G^*(\phi)$,
3. $V^*(\phi)$ is a martingale with respect to Q.

Theorem 8.2.15 *1. The price of a contingent claim X is given by $\pi(X) = E^Q[\beta_T X]$.*
2. A contingent claim is attainable $\iff V^* = V_0^* + \int H dZ$ for all H.

Definition 8.2.16 The market is called complete, if every integrable contingent claim is attainable.

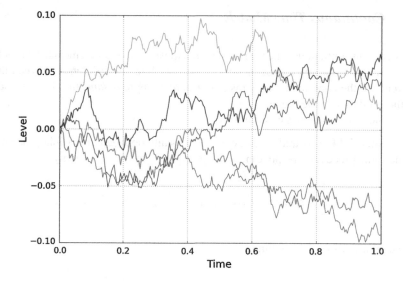

Fig. 8.3 5 Simulations of a Brownian motion

Although this theory is very important, we only gave a brief sketch of the main ideas. Thus it is recommended that the reader extends his knowledge of financial mathematics by consulting the references.

8.3 The Economic Model

As we have seen in the previous sections we need an underlying economic model to calculate the price of an option. In principle one can use various different economic models. Exemplary we are going to consider the most common model: geometric Brownian motion.

The following reference are a good sources for various aspects of the economic model: [Dot90, Duf88, Duf92, CHB89, Per94, Pli97].

Convention 8.3.1 (General conventions) *For the remainder of this chapter we will use the following notations and conventions:*

- T_x *denotes the future lifespan of an x year old person.*
- *The σ-algebras generated by T_x are denoted by $\mathcal{H}_t = \sigma\left(\{T > s\}, 0 \leq s \leq t\right).$*
- *We assume, that the values of the shares in the portfolio are given by standard Brownian motions W. (Compare with Fig. 8.3).*
- *$\mathcal{G}_t$ denotes the σ-algebra generated by W augmented by the P-null sets.*

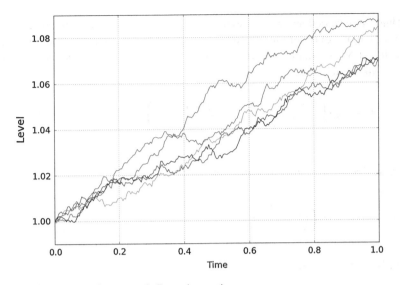

Fig. 8.4 5 Simulations of a geometric Brownian motion

Convention 8.3.2 (Independence of the financial variables)

- We assume that $\mathcal{G}_t$ and $\mathcal{H}_t$ are stochastically independent. This means, that the financial variables are independent of the future lifespan.
- $\mathcal{F}_t = \sigma\,(\mathcal{G}_t, \mathcal{H}_t)$ denotes the σ-algebra generated by $\mathcal{G}_t$ and $\mathcal{H}_t$.

Definition 8.3.3 (*Black-Scholes model*) This economic model consists of two investment options:

$$
\begin{array}{lll}
B(t) & = & \exp(\delta\,t) \\
S(t) & = & S(0)\exp\left[\left(\eta - \tfrac{1}{2}\sigma^2\right) t + \sigma\,W(t)\right]
\end{array}
\qquad
\begin{array}{l}
\text{risk free investment.} \\
\text{shares, modeled by a} \\
\text{geometric Brownian} \\
\text{motion (cf. Fig. 8.4).}
\end{array}
$$

S is the solution to the following stochastic differential equation:

$$dS = \eta\,S\,dt + \sigma\,S\,dW.$$

Exercise 8.3.4 Prove that S solves the stochastic differential equation given above.

Next we will calculate the discounted values of B and S:

$$
\begin{array}{lll}
B^*(t) = & \dfrac{B(t)}{B(t)} = & 1, \\[2ex]
S^*(t) = & \dfrac{S(t)}{B(t)} = & S(0)\exp\left[\left(\eta - \delta - \dfrac{1}{2}\sigma^2\right) t + \sigma\,W(t)\right].
\end{array}
$$

Thus we have defined the investment options. To calculate the option prices we need to find an equivalent martingale measure. That is, we have to find a measure Q such that S^* is a martingale with respect to Q. For this we define the following Radon-Nikodym density:

$$\xi_t = \exp\left(-\frac{1}{2}\left(\frac{\eta-\delta}{\sigma}\right)^2 t - \frac{\eta-\delta}{\sigma} W(t)\right) \qquad \text{for all } t \in [0, T].$$

Exercise 8.3.5 Prove the following statements:

1. $E[\xi_t] = 1$,
2. $Var[\xi_t] = \exp\left(\left(\frac{\eta-\delta}{\sigma}\right)^2 t\right) - 1$,
3. $\xi_t > 0$.

(Hint: $W(t) \sim \mathcal{N}(0, t)$.)

An application of Girsanov's theorem—a theorem in the theory of stochastic integration (e.g. [Pro90] Theorem 3.6.21)—shows that

$$\hat{W}_t = W(t) + \frac{\eta-\delta}{\sigma} t$$

is a Brownian motion with respect to $Q = \xi \cdot P$.

Naturally, after this transformation we want to prove that

$$S^*(t) = S(0) \exp\left(-\frac{1}{2}\sigma^2 t + \sigma \hat{W}(t)\right)$$

is a martingale with respect to Q. (Then the price of the option is given by its expectation with respect to Q.)

Proof We have to show the equality

$$E^Q\left[S^*(u)|\mathcal{F}_t\right] = S^*(t)$$

for $t, u \in \mathbb{R}, u > t$. With the notation $u = t + \Delta t$, $W_u = W_t + \Delta W$ and $Z \sim \mathcal{N}(0, 1)$ we have

$$
\begin{aligned}
&E^Q\left[S^*(u)|\mathcal{F}_t\right] \\
&= E^Q\left[S(0) \exp\left(-\frac{1}{2}\sigma^2 t + \sigma \hat{W}(t) + (-\frac{1}{2}\sigma^2 \Delta t + \sigma \Delta \hat{W})\right)\Big|\mathcal{F}_t\right] \\
&= S(0) \exp\left(-\frac{1}{2}\sigma^2 t + \sigma W(t)\right) E^Q\left[\exp\left(-\frac{1}{2}\sigma^2 \Delta t + \sigma\sqrt{\Delta t} Z\right)\Big|\mathcal{F}_t\right] \\
&= S^*(t).
\end{aligned}
$$

Therefore the measure Q is equivalent to P, and S^* is a martingale with respect to Q. An economist would say, "it exists (at least) one consistent price system".

Theorem 8.3.6 *Let the economic model defined above be given, i.e. it is defined by $(\Omega, \mathcal{A}, P)$, S and B. Then at time t the price of a policy with death benefit $C(T)$ is*

$$\pi_t(T) = E^Q \left[\exp\left(-\delta(T-t)\right) C(T) | \mathcal{F}_t \right].$$

Remark 8.3.7 The main difference of this model in comparison to the classical model is the fact that one has to calculate the expectation with respect to Q and not with respect to P. Moreover one should note, that we have not proved the uniqueness of the price system.

The following two formulas are an important consequence of the previous considerations.

Theorem 8.3.8 *A single premium for a policy based on the economic model defined above is given by the following formulas.*

Endowment policy:

$$V(0) = E^Q \left[\exp(-\delta T) C(T) \right] \cdot {}_T p_x.$$

Term life insurance:

$$V(0) = \int_0^T E^Q \left[\exp(-\delta t) C(t) \right] p_{**}(x, x+t) \, \mu_{*\dagger}(x+t) dt.$$

8.4 Calculation of Single Premiums

Up to now the calculations have been relatively simple, since we did not include any guarantees in our policy model. Next, we will consider a unit-linked policy with an additional guarantee. We recall some of the notations from the previous sections:

$N(\tau)$	Number of shares at time τ,
$S(\tau)$	value of a share at time τ,
$G(\tau)$	guaranteed benefits at time τ,
$C(\tau) = \max\{N(\tau)S(\tau), G(\tau)\}$	value of the insurance at time τ.

8.4.1 Pure Endowment Policy

Theorem 8.4.1 *Let the Black-Scholes model be given. Then the single premium for a pure endowment policy with payout*

$$C(T) = \max\{N(T)S(T), G(T)\}$$

is given by

$$_T G_x = {_T p_x} \left[G(T) \exp(-\delta T)\Phi\big(-d_2^0(T)\big) + S(0)N(T)\Phi\big(d_1^0(T)\big) \right],$$

where

$$\Phi(y) = \frac{1}{\sqrt{2\pi}} \int_{-\infty}^{y} \exp(-\frac{x^2}{2})dx,$$

$$d_1^t(s) = \frac{\ln\left[\frac{N(s)S(t)}{G(s)}\right] + \left(\delta + \frac{1}{2}\sigma^2\right)(s-t)}{\sigma\sqrt{s-t}}, \quad (s>t),$$

$$d_2^t(s) = \frac{\ln\left[\frac{N(s)S(t)}{G(s)}\right] + \left(\delta - \frac{1}{2}\sigma^2\right)(s-t)}{\sigma\sqrt{s-t}}, \quad (s>t).$$

Proof In the following we denote by J^* the discounted value of a random variable J. The value of the pure endowment policy at time zero is $E^Q[C^*(T)]$. We set $Z = S^*(T)$. Then the following equations hold

$$_T G_x = {_T p_x}\, E^Q\left[\max\left\{N(T)Z, G^*(T)\right\}\right]$$

and

$$Z = S(0)\exp\left(-\frac{1}{2}\sigma^2 T + \sigma\hat{W}(T)\right) \quad \text{where} \quad \hat{W}(T) \sim \mathcal{N}(0, T).$$

Thus we get

$$_T G_x = {_T p_x} \int_{-\infty}^{\infty} \max\left[N(T)S(0)\exp(-\frac{1}{2}\sigma^2 T + \sigma\xi), G^*(T)\right] f(\xi)d\xi,$$

$$f(\xi) = \frac{1}{\sqrt{2\pi T}} \exp\left(-\frac{1}{2T}\xi^2\right).$$

Next we define $\bar{\xi} = \frac{1}{\sigma}\left[\ln\left(\frac{G^*(T)}{N(T)S(0)}\right) + \frac{1}{2}\sigma^2 T\right]$ and note that $\xi > \bar{\xi}$ implies $N(T)Z > G^*(T)$. Therefore, the single premium is given by

$$_T G_x = {_T p_x} \left(G^*(T) \int_{-\infty}^{\bar{\xi}} f(\xi) d\xi \right.$$

$$+ N(T) S(0) \int_{\bar{\xi}}^{\infty} \exp(-\frac{1}{2}\sigma^2 T + \sigma \xi) f(\xi) d\xi \right)$$

$$= {_T p_x} \left(G^*(T) \int_{-\infty}^{\bar{\xi}} f(\xi) d\xi \right.$$

$$+ N(T) S(0) \int_{\bar{\xi}}^{\infty} \frac{1}{\sqrt{2\pi T}} \exp(-\frac{1}{2T}(\xi - \sigma T)^2 d\xi \right).$$

This equation, with adapted notation, yields the statement of the theorem.

8.4.2 Term Life Insurance

Theorem 8.4.2 *Let the Black-Scholes model be given. Then the single premium for a term life insurance with death benefit*

$$C(t) = \max\{N(t)S(t), G(t)\}$$

is given by

$$G^1_{x:T} = \int_0^T \left(G(t) \exp(-\delta t) \Phi(-d_2^0(t)) + S(0) N(t) \Phi(d_1^0(t))_t \, p_x \mu_{x+t} \right) dt,$$

where

$$\Phi(y) = \int_{-\infty}^y \frac{1}{\sqrt{2\pi}} \exp(-\frac{x^2}{2}) dx,$$

$$d_1^t(s) = \frac{\ln\left[\frac{N(s)S(t)}{G(s)}\right] + (\delta + \frac{1}{2}\sigma^2)(s - t)}{\sigma\sqrt{s - t}},$$

$$d_2^t(s) = \frac{\ln\left[\frac{N(s)S(t)}{G(s)}\right] + (\delta - \frac{1}{2}\sigma^2)(s - t)}{\sigma\sqrt{s - t}},$$

for s > t.

Exercise 8.4.3 Prove the previous theorem by the same methods which we used for the pure endowment policy.

8.5 Thiele's Differential Equation

Now we want to derive Thiele's differential equation. For this we need to determine premiums for the policies. We introduce the notation $\bar{p}(t)$ for the density of the premiums at time t. Then the equivalence principle yields the following two equations:

$$_T G_x = \int_0^T \bar{p}(t) \exp(-\delta t)\, _t p_x dt$$

and

$$G_{x:T}^1 = \int_0^T \bar{p}(t) \exp(-\delta t)\, _t p_x dt.$$

Also in this section the pure endowment policy and the term life insurance will be considered separately. The mathematical reserve for these policies is given by:

Pure endowment: $V(t) = {}_{T-t} p_{x+t} \pi_t(T)$
$$-\int_t^T \bar{p}(\xi) \exp(-\delta(\xi - t))\, _{\xi-t} p_{x+t}\, d\xi.$$

Life insurance: $V(t) = \int_t^T \left(\pi_t(\xi)\mu_{x+\xi} - \bar{p}(\xi) \exp(-\delta(\xi - t)) \right)$
$$\times _{\xi-t} p_{x+t}\, d\xi,$$

where

$$\pi_t(s) = G(s) \exp(-\delta(s - t))\Phi(-d_2^t(s))$$
$$+ N(s)S(t)\Phi(d_1^t(s)),$$

$$d_1^t(s) \;\; = \;\; \frac{\ln\left[\frac{N(s)S(t)}{G(s)}\right] + \left(\delta + \frac{1}{2}\sigma^2\right)(s - t)}{\sigma\sqrt{s - t}},$$

$$d_2^t(s) \;\; = \;\; \frac{\ln\left[\frac{N(s)S(t)}{G(s)}\right] + \left(\delta - \frac{1}{2}\sigma^2\right)(s - t)}{\sigma\sqrt{s - t}},$$

for $s > t$.

Remark 8.5.1 • In the classical setting the reserves were deterministic, but here they depend on the underlying share S.
• Note that we are beyond the deterministic theory of differential equations. In particular we have to use Itô's formula, which takes the following form for the purely continuous case of a standard Brownian motion W:

$$df(W) = f'\,dW + \frac{1}{2}f''\,ds.$$

For the policies defined above we have the following theorem.

Theorem 8.5.2 *1. The differential equation for the price of a pure endowment policy is:*

$$\frac{\partial V}{\partial t} = \bar{p}(t) + (\mu_{x+t} + \delta)\,V(t) - \frac{1}{2}\sigma^2 S(t)^2 \frac{\partial^2 V}{\partial S^2} - \delta\,S(t)\frac{\partial V}{\partial S}.$$

2. The differential equation for the price of a term life insurance is:

$$\frac{\partial V}{\partial t} = \bar{p}(t) + (\mu_{x+t} + \delta)\,V(t) - C(t)\mu_{x+t} - \frac{1}{2}\sigma^2 S(t)^2 \frac{\partial^2 V}{\partial S^2} - \delta\,S(t)\frac{\partial V}{\partial S}.$$

Before we prove this theorem, we want to make some comments on the formulas.

Remark 8.5.3 1. One obtains Black-Scholes formula by setting $\mu_{x+t} = \bar{p}(t) = 0\,\forall t$.
2. The first terms in the differential equations in the theorem above coincide with the classical case, i.e. the dependence of the values on the premiums, on the mortality and on the interest rate. Due to the shares in the model a further term appears: $-\frac{1}{2}\sigma^2 S(t)^2 \frac{\partial^2 V}{\partial S^2} - \delta\,S(t)\frac{\partial V}{\partial S}$. It represents the fluctuations of the underlying shares.

Proof We have

$$\pi_t^*(T) = \exp(-\delta t)\pi_t(T).$$

Hence, by the definition of V, we get

$$V(t) = {}_{T-t}p_{x+t}\pi_t^*(T)\exp(\delta t) - \int_t^T \bar{p}(\xi)\exp(-\delta(\xi - t))_{\xi-t}p_{x+t}d\xi$$

and

$$\pi_t^*(T) = \Psi(t)\left[V(t) + \int_t^T \bar{p}(\xi)\exp(-\delta(\xi - t))_{\xi-t}p_{x+t}d\xi\right],$$

where

$$\Psi(t) = \frac{\exp(-\delta t)}{{}_{T-t}p_{x+t}}.$$

Now we can apply Itô's formula to the function $\pi_t^*(t, S)$, since π_t^* is a function of S and t. We get

$$dY_t = U(t + dt, X_t + dX_t) - U(t, X_t)$$
$$= \left(U_t dt + \frac{1}{2} U_{xx} b^2 dt \right) + U_x dX_t$$
$$= \left(U_t + \frac{1}{2} U_{xx} b^2 \right) dt + U_x\, b\, dB_t$$

and

$$d\pi^* = \left(\frac{\partial \pi^*}{\partial t} + \frac{\partial \pi^*}{\partial S} a + \frac{1}{2} \frac{\partial^2 \pi^*}{\partial S^2} b^2 \right) dt + \frac{\partial \pi^*}{\partial S} b\, d\hat{W}.$$

Furthermore we know that

$$dS = \delta S(t) dt + \sigma S(t) d\hat{W},$$

and thus we have $a = \delta S(t)$ and $b = \sigma S(t)$. In the next step we want to determine the two terms:

$$\frac{\partial \pi^*_t}{\partial S} = \Psi(t) \frac{\partial V}{\partial S},$$
$$\frac{\partial^2 \pi^*_t}{\partial S^2} = \Psi(t) \frac{\partial^2 V}{\partial S^2}.$$

To get $\frac{\partial \pi^*}{\partial t}$, we start with

$$\frac{\partial}{\partial t} {}_{\xi-t} p_{x+t} = \mu_{x+t} \, {}_{\xi-t} p_{x+t},$$

$$\frac{\partial}{\partial t} \Psi(t) = \left(\frac{A}{B} \right)' = \frac{A'}{B} - \frac{A}{B^2} B'$$
$$= -(\mu_{x+t} + \delta)\, \Psi(t).$$

Now, with the formula from above we get

$$\frac{\partial \pi^*}{\partial t} = \frac{\partial \Psi}{\partial t} \left(V(t) + \int_t^T \bar{p}(\xi) \exp(-\delta(\xi - t))_{\xi-t} p_{x+t} dt \right)$$
$$+ \Psi(t) \left(\frac{\partial V}{\partial t} + \frac{\partial}{\partial t} \int_t^T \bar{p}(\xi) \exp(-\delta(\xi - t))_{\xi-t} p_{x+t} dt \right)$$
$$= \Psi(t) \left(\frac{\partial V}{\partial t} - (\mu_{x+t} + \delta) V(t) - \bar{p}(t) \right),$$

where we applied the chain rule to

$$\frac{\partial}{\partial t} \int_t^T \bar{p}(\xi) \exp(-\delta(\xi - t))_{\xi-t} p_{x+t} dt.$$

Thus we get

$$\pi_s^*(T) = \pi_t^*(T) + \int_t^s \Psi(\xi) \frac{\partial V}{\partial S} \sigma S d\hat{W}(\xi)$$

$$+ \int_t^s \Psi(\xi) \left[\frac{\partial V}{\partial S} \delta S + \frac{1}{2}\sigma^2 S^2 \frac{\partial^2 V}{\partial S^2} - (\mu_{x+\xi} + \delta)V(\xi) \right.$$

$$\left. + \frac{\partial V}{\partial t}(\xi)\bar{p}(\xi) \right] d\xi.$$

Now the drift term is equal to zero, since $\pi_\cdot^*(T)$ is a martingale. Therefore we finally get

$$\frac{\partial V}{\partial t} = \bar{p}(t) + (\mu_{x+t} + \delta) V(t) - \frac{1}{2}\sigma^2 S(t)^2 \frac{\partial^2 V}{\partial S^2} - \delta S(t) \frac{\partial V}{\partial S}.$$

Exercise 8.5.4 Prove the second part of the theorem above.

Chapter 9
Policies with Stochastic Interest Rate

9.1 Introduction

In the previous chapter we looked at unit-linked policies. Based on these we will now consider policies with a technical interest rate modelled by a stochastic process, which is for example given by a stochastic differential equation.

As before we will assume that the random variables of the economic model are independent of the random variables which describe the state of the insured. We also assume that the market does not provide arbitrage opportunities. Thus one has to find an equivalent martingale measure to price the option which models the policy. The price of the option is then given as expectation with respect to this new measure.

Definition 9.1.1 (*Spotrate*) In the following we will denote the spot interest rate by r_t, this is the interest rate which is currently paid. The accumulated interest for an interval $[0, t]$ is given by

$$\gamma_t := \beta_t^{-1} = \exp\left(\int_0^t r_s ds\right),$$

which is the solution to the stochastic differential equation

$$d\gamma_t = r_t \gamma_t \, dt,$$

with $\gamma_0 = 1$.

The value of a zero coupon bond at time t, paying 1 at time s, is denoted by $B_t(s)$.

9.2 The Vasicek Model

In the following we look at a particular interest rate model, the Vasiček model. It is defined by the stochastic differential equation

M. Koller, *Stochastic Models in Life Insurance*, EAA Series,
DOI: 10.1007/978-3-642-28439-7_9, © Springer-Verlag Berlin Heidelberg 2012

$$dr_t = \alpha \, (\rho - r_t) \, dt + \sigma \, dW_t, \tag{9.1}$$

where $\alpha > 0$ and $\rho, \sigma \in \mathbb{R}$. Thus, the interest rate process is induced by a Brownian motion W. The model has the feature, that in the absence of stochastic noise the interest rate would converge towards ρ (mean reversion). The stochastic process given by (9.1) is also known as Ornstein-Uhlenbeck process.

We chose the Vasiček model as primary example, since for this model it is possible to calculate the quantities relevant to us (e.g. the discount rate) explicitly. Thus one can use the values given by this model directly. Moreover, it is easily possible to extend the model by replacing the Brownian motion W_t by another stochastic process [Nor98]. A shortcoming of this model is the fact, that it allows with positive probability negative interest rates.

Theorem 9.2.1 *The following statements hold for the Vasiček model:*

1. $r_t \sim \mathcal{N}\left(\rho + \exp(-\alpha t)\,(r_0 - \rho),\, \frac{\sigma^2}{2\alpha}\,(1 - \exp(-2\alpha t))\right),$
2. $Cov(r_s, r_t) = \exp(-\alpha(s+t))\,\frac{\sigma^2}{2\alpha}\,(\exp(2\alpha t) - 1),\, for\ s \leq t.$

Proof [Nor98, Pro90, KS88].

Exercise 9.2.2 Calculate a 95% confidence interval for r_t with $\alpha = 0.1$, $\rho = 0.05$, $r_0 = 0.03$ and $\sigma = 0.01$ for a period of 20 years.

Next, we have a look at the accumulated interest $y(t) = \exp(\int_0^t r_s \, ds)$ on the interval $[0, t]$.

Theorem 9.2.3 *The following equations hold for the Vasiček model:*

1. $E[y(t)] = \rho t + (r_0 - \rho)\frac{1 - \exp(-\alpha t)}{\alpha},$
2. $Var[y(t)] = \frac{\sigma^2}{\alpha^2}t + \frac{\sigma^2}{2\alpha^3}\left[-3 + 4\exp(-\alpha t) - \exp(-2\alpha t)\right],$
3. $Cov(y(s), y(t)) = \frac{\sigma^2}{\alpha^2}\min(s,t) + \frac{\sigma^2}{2\alpha^3}\left[-2 + 2\exp(-\alpha s)\right.$

$\left. +2\exp(-\alpha t) - \exp(-\alpha|t - s|) - \exp(-\alpha(t + s))\right].$

Proof This theorem is a consequence of Theorem 9.2.1. One just has to change the order of integration, which is allowed by Fubini's theorem.

Exercise 9.2.4 Complete the proof of the theorem above.

9.3 Value of the Portfolio

In this section we will analyse the influence of the stochastic interest rate on the value of the portfolio. For insurance policies with deterministic interest rate we know that the present values of the cash flows of the policies are independent and their joint variance $V_{Tot} = \sum_{k=1}^n V_k$ converges with rate $1/n$ to zero (central limit theorem).

This does not hold, if the interest rate is a stochastic process. In this case the present values are dependent, since the same interest rate process is used for all policies. In the following we want to concentrate on the most important aspects of this setup. Therefore we make the following assumptions:

- We work with a discrete time model. Thus, we do not have to consider double integrals.
- We assume that the state in which benefits are payed to the insured is absorbing. Furthermore we also assume that the insured event can only occur at the end of each period. These assumptions are very common for pensions or life insurances or their combination in an endowment policy.

In order to be able to derive some results, we start with setting some notations.

Definition 9.3.1 In the following Z denotes the present value of the future benefits discounted to time 0. The time of the occurrence of the insured event is denoted by K. Moreover we assume, that a payment b_k is due at time $t(k)$, if the insured event occurs at time $K = k$.

As usual, we denote the probability to die within t years (to survive t years, respectively) for an x year old person by

$$_tq_x = P[K \leq t],$$
$$_tp_x = 1 - {}_tq_x,$$

respectively. In the following we will always consider "death" as the insured event, even if in the concrete example another event is insured.

Theorem 9.3.2 *With the notation introduce above the following equation holds:*

$$A := E[Z] = \sum_{k=0}^{\infty} b_k E[\exp(-y(t(k)))]_k p_x q_{x+k},$$

where $\exp(-y(t)) = \exp(-\int_0^t r_s ds)$ *denotes the (stochastic) discount from time* t *to time* 0.

Proof This identity is a direct consequence of the previous definitions and the tower property of the expectation (i.e., $E[Z] = E[E[Z \mid K]]$).

In the same manner one can calculate moments of higher order of Z:

Theorem 9.3.3 *Let* $m \in \mathbb{N}$. *Then the m-th moment of the present value of Z is*

$$E[Z^m] = \sum_{k=0}^{\infty} b_k^m E[\exp(-\tilde{y}(t(k)))]_k p_x q_{x+k},$$

$$\text{where } \tilde{y}(t) = \int_0^t m\, r_\tau d\tau$$

denotes the accumulated interest for the m-th power of the interest intensity. In the case of an insurance whose payouts are either 0 or 1 the previous equation simplifies to

$$E[Z^m] = \sum_{k=0}^{\infty} b_k \, E[\exp(-\tilde{y}(t(k)))]_k p_x q_{x+k}.$$

Proof These equations can be obtained by calculating the conditional expectation $E[Z^m \mid K]$:

$$E[Z^m] = E[E[Z^m \mid K]]$$

$$= \sum_{k=0}^{\infty} E[Z^m \mid K = k]_k p_x q_{x+k}$$

$$= \sum_{k=0}^{\infty} E[\{b_k \, \exp(-y(t(k)))\}^m]_k p_x q_{x+k}$$

$$= \sum_{k=0}^{\infty} b_k^m \, E[\exp(-m \, y(t(k)))]_k p_x q_{x+k}$$

$$= \sum_{k=0}^{\infty} b_k^m \, E[\exp(-\tilde{y}(t(k)))]_k p_x q_{x+k}.$$

The second equation is implied by the facts, that $1^m = 1$ und $0^m = 0$.

Definition 9.3.4 (*Portfolio assumptions*) We consider a portfolio with c identical policies, and we use the notations: nt value of the future benefits of policy i. The

> K_i time of the occurrence of the insured event for policy i,
> Z_i prese

sum of the present values of all c policies is

$$Z(c) := \sum_{k=1}^{c} Z_i.$$

The aim of the next part is to analyse Z. For this we have to make certain assumptions about our portfolio:

Remark 9.3.5 Note that the present values are not independent, even though the conditional present values are independent. This is due to the fact, that all present values are subject to the same discount $\exp(-y(t))$.

A1 The times of the occurrences of the insured events are independent and identical distributed.

A2 The random variables of the economic model are independent of the times of the occurrences of the insured events. In other words: family $\{K_i : i = 1, ...c\}$ is independent of the family $\{\delta_\tau : \tau > 0\}$.

A3 All policies have an identical structure, i.e. the conditional present values $\{Z_i \mid (y_n)_{n \in \mathbb{N}}\}_{i=1,...c}$ are independent and identical distributed.

We begin with a calculation of the present values of the whole portfolio.

Theorem 9.3.6 *Let the assumptions A1, A2 and A3 hold. Then the present value of the portfolio is given by*

$$E[Z(c)] = c\, E[Z_1].$$

Proof

$$E[Z(c)] = E[\sum_{k=1}^{c} Z_i] = \sum_{k=1}^{c} E[Z_i] = c\, E[Z_1].$$

Analogous, one can also calculate the second moment.

Theorem 9.3.7 *Let the assumptions A1, A2 and A3 hold. Then the second moment of the present value of the portfolio is*

$$E[Z(c)^2] = c(c-1)E[Z_1 Z_2] + c\, E[Z_1^2],$$

where

$$E[Z_1 Z_2] = \sum_{i=0}^{\infty} \sum_{j=0}^{\infty} b_i\, b_j\, E[\exp(-y(t(i)) - y(t(j)))]_i\, {}_i p_x q_{x+i}\, {}_j p_x q_{x+j}.$$

Proof The theorem is proved by a direct verification of the formula

$$E[Z(c)^2] = E[(\sum_{i=1}^{c} Z_i)^2]$$

$$= \sum_{i,j=1}^{c} E[Z_i Z_j]$$

$$= \sum_{i=1}^{c} E[Z_i^2] + 2\sum_{i<j} E[Z_i Z_j].$$

Now each part can be calculated. We get

$$
\begin{aligned}
E[Z_i^2] &= E[E[Z_i^2 \mid (y_n)_{n\in\mathbb{N}}]] \\
&= E[E[Z_1^2 \mid (y_n)_{n\in\mathbb{N}}]] \\
&= E[Z_1^2]
\end{aligned}
$$

and similarly for $i \neq j$

$$
\begin{aligned}
E[Z_i\, Z_j] &= E[E[Z_i\, Z_j \mid (y_n)_{n\in\mathbb{N}}]] \\
&= E[E[Z_i \mid (y_n)_{n\in\mathbb{N}}] \times E[Z_j \mid (y_n)_{n\in\mathbb{N}}]] \\
&= E[E[Z_1 \mid (y_n)_{n\in\mathbb{N}}] \times E[Z_2 \mid (y_n)_{n\in\mathbb{N}}]] \\
&= E[Z_1\, Z_2].
\end{aligned}
$$

Replacing these terms in the above equation yields the statement of the theorem.

The second moment of the portfolio can be used to calculate its asymptotic variance.

Theorem 9.3.8 *The asymptotic variance satisfies the equation:*

$$
\lim_{c\to\infty} Var(\frac{Z(c)}{c}) = E[Z_1\, Z_2] - E[Z_1]^2.
$$

Proof The following identities hold:

$$
\begin{aligned}
Var\left(\frac{Z(c)}{c}\right) &= E[(\frac{Z(c)}{c})^2] - E[\frac{Z(c)}{c}]^2 \\
&= \frac{c(c-1)}{c^2} E[Z_1\, Z_2] + \frac{c}{c^2} E[Z_1^2] - E[Z_1]^2 \\
&\to E[Z_1\, Z_2] - E[Z_1]^2 \text{ for } c \to \infty.
\end{aligned}
$$

Remark 9.3.9 Contrary to the classical model with deterministic interest rate, here the asymptotic variance (for $c \to \infty$) is not equal to zero. This shows, that a stochastic oscillation of the bonds induces a positive asymptotic variance. Furthermore, this risk can not be reduced by raising the number of policies in the portfolio.

Theorem 9.3.10 *Let assumptions A1, A2 and A3 hold. Then the third moment of the present value of the portfolio is*

$$
\begin{aligned}
E[Z(c)^3] &= c(c-1)(c-2)\; E[Z_1\, Z_2\, Z_3] \\
&\quad + c(c-1)\; E[Z_1^2\, Z_2] \\
&\quad + c\; E[Z_1^3],
\end{aligned}
$$

where

$$E[Z_1 \, Z_2 \, Z_3] = \sum_{i=0}^{\infty} \sum_{j=0}^{\infty} \sum_{k=0}^{\infty} b_i \, b_j \, b_k \, {}_i p_x q_{x+i} \, {}_j p_x q_{x+j} \, {}_k p_x q_{x+k}$$
$$\times E[\exp(-y(t(i)) - y(t(j)) - y(t(k)))],$$

$$E[Z_1^2 \, Z_2] = \sum_{i=0}^{\infty} \sum_{j=0}^{\infty} b_i^2 \, b_j \, {}_i p_x q_{x+i} \, {}_j p_x q_{x+j} E[\exp(-2y(t(i)) - y(t(j)))],$$

$$E[Z_1^3] = \sum_{i=0}^{\infty} b_i^3 \, {}_i p_x q_{x+i} E[\exp(-3y(t(i)))].$$

Proof The proof is analogous to the proof of the formula for the second moment. One starts by decomposing $E[Z(c)^3]$ into

$$E[Z(c)^3] = \sum_{i,j,k=1}^{c} E[Z_i \, Z_j \, Z_k].$$

Then the terms are sorted based on their structure, and one proves the following identities:

$$E[Z_i \, Z_j \, Z_k] = E[Z_1 \, Z_2 \, Z_3], \text{ if } i \neq j \neq k \neq i,$$
$$E[Z_i^2 \, Z_j] = E[Z_1^2 \, Z_2], \text{ if } i \neq j.$$

Using these identities in the decomposition yields the statement of the theorem.

Theorem 9.3.11 *The formula of Theorem 9.3.10 can also be used to calculate the asymptotic skewness* $Sk(Z) := \frac{E[(Z-E[Z])^3]}{Var(Z)^{3/2}}$. *We have*

$$\lim_{c \to \infty} Sk(Z(c)) = \frac{E[Z_1 \, Z_2 \, Z_3] - 3E[Z_1 \, Z_2] \times E[Z_1] + 2E[Z_1]^3}{(E[Z_1 \, Z_2] - E[Z_1]^2)^{3/2}}.$$

Proof To prove this theorem we set $W = \frac{Z(c)}{c}$. Then the following identities hold:

$$Sk(Z) = \frac{E[(Z-E[Z])^3]}{Var(Z)^{3/2}}$$
$$= \frac{E[W^3 - 3W^2 E[W] + 3W E[W]^2 - E[W]^3]}{(Var(W))^{3/2}}$$
$$= \frac{1}{(Var(W))^{3/2}} \times \left[\frac{c(c-1)(c-2)}{c^3} E[Z_1 Z_2 Z_3] + \frac{c(c-1)}{c^3} E[Z_1^2 Z_2] \right.$$

$$+ \frac{c}{c^3} E[Z_1^3] - 3 \frac{cE[Z_1]}{c} \left\{ \frac{c(c-1)}{c^2} E[Z_1 Z_2] + \frac{c}{c^2} E[Z_1^2] \right\}$$

$$+ 3 \frac{c^2 E[Z_1]^2}{c^2} \frac{cE[Z_1]}{c} - \left[\frac{cE[Z_1]}{c} \right]^3 \Bigg]$$

$$\rightarrow \frac{E[Z_1 Z_2 Z_3] - 3E[Z_1 Z_2] \times E[Z_1] + 2E[Z_1]^3}{(E[Z_1 Z_2] - E[Z_1]^2)^{3/2}}, \quad \text{for } c \rightarrow \infty.$$

The previous formulas are all based on direct calculations of the values in question. This is in contrast to the recursion techniques for the Markov model, where we based statements for time n on statements for time $n - 1$.

In order to derive also recursion equations in the current setting we have to make further assumptions.

Definition 9.3.12 (*Portfolio assumptions*) We consider a portfolio with c identical policies with state space S, and we use the following notations:

$_i X$ the Markov chain on the state space S, which represents the state of the i-th policy,

W the Markov chain on the state space $\tilde{S}$, which determines the bond process. Here, $v_t = \sum_{\iota \in \tilde{S}} I_\iota^W(t) v_\iota(t)$ holds.

These notations and the following definitions will enable us to derive recursion formulas for policies modelled by Markov chains.

We set the following notations:

$$V_{il}^\alpha(t) = E[_1 V^{+,\alpha} |_1 X = i, \ W = l],$$

$$V_{ijl}^{\alpha\beta}(t) = E[_1 V^{+,\alpha} \, _2 V^{+,\beta} |_1 X = i, \ _2 X = j, \ W = l],$$

$$V_{ijkl}^{\alpha\beta\gamma}(t) = E[_1 V^{+,\alpha} \, _2 V^{+,\beta} \, _3 V^{+,\gamma} |_1 X = i, \ _2 X = j, \ _3 X = k, \ W = l],$$

where $_1 V^{+,\alpha}$ denotes α-th power of the prospective reserve of the first policy. The sum of all present values of all c policies is

$$Z(c) := \sum_{k=1}^{c} {_k V^+}.$$

The aim of the next part is to analyse Z. For this we have to make certain assumptions about our portfolio:

A1   The processes $\{_i X | i = 1, \ldots, c\}$ are independent and identical distributed.

A2   The random variables of the economic model are independent of the times of the occurrences of the insured events.

A3   All policies have an identical structure, i.e. the conditional present values $\{_i V^+ \mid (y_n)_{n \in \mathbb{N}}, {}_i X = \iota\}_{i=1,\ldots c}$ are independent and identical distributed.

Under these assumptions on the portfolio we are able to derive several recursion formulas.

Theorem 9.3.13 *In the above setting the following formulas hold:*

$$E[Z(c)|\{_\kappa X(t) = i : \kappa = 1 \ldots c, W(t) = w\}] = c\, E[V_1|_1 X(t) = i, W(t) = w],$$

$$
\begin{aligned}
&E[Z(c)^2|\{_\kappa X(t) = i : \kappa = 1, \ldots, c, W(t) = w\}] \\
&\quad = cE[_1 V^2|_1 X(t) = i, W(t) = w] \\
&\qquad + c(c-1)E[_1 V\, _2 V|_1 X(t) = i, _2 X(t) = i, W(t) = w], \\
&E[Z(c)^3|\{_\kappa X(t) = i : \kappa = 1, \ldots, c, W(t) = w\}] \\
&\quad = c(c-1)(c-2)E[_1 V\, _2 V\, _3 V|_1 X(t) = i, _2 X(t) = i, _3 X(t) = i, W(t) = w] \\
&\qquad + c(c-1)E[_1 V^2\, _2 V|_1 X(t) = i, _2 X(t) = i, W(t) = w] \\
&\qquad + c\, E[_1 V^3|_1 X(t) = i, W(t) = w].
\end{aligned}
$$

Proof The proof is analogous to the proofs of Theorems 9.3.7 and 9.3.10.

Now we have seen that recursion formulas also hold in this setting. Next, we want to calculate explicitly the terms appearing therein. By algebraic transformation, values at time t will be calculated based on the values at time $t + 1$. We start with the calculation of $V_{ijl}^{11}(t)$.

Theorem 9.3.14 *In the discrete time Markov model the prospective reserves $V_{ijl}^{11}(t)$ with $i, j \in S$ and $l \in \tilde{S}$ satisfy the following recursion:*

$$
\begin{aligned}
V_{ijl}^{11}(t) = &\sum_{\bar{i},\bar{j},\bar{l}} p_{i\bar{i}}(t, t+1)\, p_{j\bar{j}}(t, t+1)\, p_{l\bar{l}}^W(t, t+1)\, v_{\bar{l}}^2\, V_{\bar{i},\bar{j},\bar{l}}^{11}(t+1) \\
&+ \sum_{\bar{i},\bar{j},\bar{l}} p_{i\bar{i}}(t, t+1)\, p_{j\bar{j}}(t, t+1)\, p_{l\bar{l}}^W(t, t+1)\, v_{\bar{l}}\, V_{\bar{i},\bar{l}}^{1}(t+1)\, \Theta_{j\bar{j}l}(t) \\
&+ \sum_{\bar{i},\bar{j},\bar{l}} p_{i\bar{i}}(t, t+1)\, p_{j\bar{j}}(t, t+1)\, p_{l\bar{l}}^W(t, t+1)\, v_{\bar{l}}\, V_{\bar{j},\bar{l}}^{1}(t+1)\, \Theta_{i\bar{i}l}(t)
\end{aligned}
$$

$$+ \sum_{\tilde{i},\tilde{j},\tilde{l}} p_{i\tilde{i}}(t,t+1)\, p_{j\tilde{j}}(t,t+1)\, p_{l\tilde{l}}^{W}(t,t+1)\, \Theta_{\tilde{i}\tilde{i}l}(t)\, \Theta_{\tilde{j}\tilde{j}l}(t),$$

where we used the notation

$$\Theta_{ijk}(t) = a_i^{Pre}(t) + v_k(t)\, a_{ij}^{Post}(t).$$

Proof The proof is based on a general principle: it considers on the one hand the current period and on the other hand the future periods. Thereby the sum is decomposed into two parts, which leads to the result.

We want to calculate the quantity $E[_1V^1\,_2V^1|_1X_t = i,\,_2X_t = j,\,W_t = l]$. First we get the formula

$$_1V^1 = \sum_{k=0}^{\infty}\left(\prod_{l=0}^{l<k} v_l\right)\left(\sum_{\iota\in I} I_\iota(k)a_\iota^{Pre}(k) + v_k \sum_{\iota,\kappa\in I}\Delta N_{\iota\kappa}(k)a_{\iota\kappa}^{Post}(k)\right).$$

Similarly, one can also calculate $_2V^1$. In a next step this formula is transformed to

$$_1V^1 = \sum_{k=0}^{\infty}\left(\prod_{l=0}^{l<k} v_l\right)\times\left(\sum_{\iota\kappa\in I}\Delta N_{\iota\kappa}(k)\Theta_{\iota\kappa\bullet}(k)\right).$$

Now $_1V^1 \times _2V^1$ can be calculated. For this we denote the term in the first sum by $\Upsilon^\iota(k)$. Thus the formula above simplifies to

$$_1V^1 = \sum_{k=0}^{\infty}\Upsilon^1(k), \tag{9.2}$$

and we can do the following calculation:

$$_1V^1(x)\times _2V^1(x) = \left(\sum_{k_1=x}^{\infty}\Upsilon^1(k_1)\right)\times\left(\sum_{k_2=x}^{\infty}\Upsilon^2(k_2)\right)$$

$$= \left(\Upsilon^1(x) + \sum_{k_1=x+1}^{\infty}\Upsilon^1(k_1)\right)\times\left(\Upsilon^2(x) + \sum_{k_2=x+1}^{\infty}\Upsilon^2(k_2)\right)$$

$$= (A_1 + A_2)(B_1 + B_2)$$

$$= A_1 B_1 + A_2 B_1 + A_1 B_2 + A_2 B_2.$$

Finally, taking expectations of these terms yields the statement of the theorem.

The above calculation shows that one could apply the same method to moments of arbitrary order, just the formulas would become more involved. For example one can prove the following.

Theorem 9.3.15 *The following recursion formulas hold:*

$$V_{ij\bar{l}}^{12}(t) = \sum_{\bar{i},\bar{j},\bar{l}} p_{i\bar{i}}(t, t+1) \, p_{j\bar{j}}(t, t+1) \, p_{l\bar{l}}^{W}(t, t+1) \, \Theta_{\bar{i}\bar{i}l}(t) \Theta_{j\bar{j}l}(t)^2$$

$$+ \sum_{\bar{i},\bar{j},\bar{l}} p_{i\bar{i}}(t, t+1) \, p_{j\bar{j}}(t, t+1) \, p_{l\bar{l}}^{W}(t, t+1) \, v_l \, V_{\bar{i},\bar{i}}^{1}(t+1) \Theta_{j\bar{j}l}(t)^2$$

$$+ 2 \sum_{\bar{i},\bar{j},\bar{l}} p_{i\bar{i}}(t, t+1) p_{j\bar{j}}(t, t+1) p_{l\bar{l}}^{W}(t, t+1) \, v_l \, V_{\bar{j},\bar{l}}^{1}(t+1) \Theta_{\bar{i}\bar{i}l}(t) \Theta_{j\bar{j}l}(t)$$

$$+ 2 \sum_{\bar{i},\bar{j},\bar{l}} p_{i\bar{i}}(t, t+1) p_{j\bar{j}}(t, t+1) p_{l\bar{l}}^{W}(t, t+1) v_l^2 V_{\bar{i},\bar{i}}^{1}(t+1) V_{\bar{j},\bar{l}}^{1}(t+1) \Theta_{j\bar{j}l}(t)$$

$$+ \sum_{\bar{i},\bar{j},\bar{l}} p_{i\bar{i}}(t, t+1) \, p_{j\bar{j}}(t, t+1) \, p_{l\bar{l}}^{W}(t, t+1) \, v_l^2 \, V_{\bar{j},\bar{l}}^{2}(t+1) \Theta_{j\bar{j}l}(t)$$

$$+ \sum_{\bar{i},\bar{j},\bar{l}} p_{i\bar{i}}(t, t+1) \, p_{j\bar{j}}(t, t+1) \, p_{l\bar{l}}^{W}(t, t+1) \, v_l^3 \, V_{\bar{i}\bar{j},\bar{l}}^{12}(t+1),$$

where we used the notation

$$\Theta_{ijk}(t) = a_i^{Pre}(t) + v_k(t) a_{ij}^{Post}(t).$$

Exercise 9.3.16 Prove the previous theorem and try to find a general formula for the derivation of the equation given above. For this it is in each case necessary to decompose the sum (9.2) in a part corresponding to the benefits which are due in the current period and in a part for the future periods. For example, if we want to calculate $E\left[XY^2Z^3\right]$, we have to consider

$$(X_1 + X_2)(Y_1 + Y_2)^2(Z_1 + Z_2)^3,$$

where $\bullet_1$ denotes the payments in the current period and $\bullet_2$ denotes the future payments of the corresponding process. How does the recursion formula for $V_{ijkl}^{111}(t)$ look like?

9.4 A Model for the Interest Rate Structure

In the previous section we have seen, how a stochastic interest rate influences a portfolio. In this section we will focus on a concrete interest rate model. The interest rate, which we are going to use, is given by the following stochastic differential equation:

$$r_t = r_0 + \int_0^t \eta_t(r_s, s)ds + \int_0^t \sigma(r_s, s)dW,$$

where r_0 denotes the spot rate at time 0, and we assumed that η and σ are of sufficient regularity. Furthermore we assume, that $\hat{Z}$ is an $n - 1$ dimensional Itô process. We set $Z = \begin{pmatrix} r \\ \hat{Z} \end{pmatrix}$ for

$$Z_t = Z_0 + \int_0^t \eta_Z(Z_s, s)ds + \int_0^t \sigma_Z(Z_s, s)ds,$$

where $\eta_Z : \mathbb{R}^n \to \mathbb{R}^n$ and $\sigma_Z : \mathbb{R}^n \to \mathbb{R}^{n \times d}$. Finally we assume that $B_t(s) = B_s(Z_t, t)$ for a B of sufficient regularity.

In this setting the following theorem holds.

Theorem 9.4.1 B satisfies the stochastic differential equation

$$B_t(s) = B_0(s) + \int_0^t \eta_B(u, s)B_u(s)du + \int_0^t B_u(s)\sigma_B dW_u,$$

where

$$\eta_B(t, s) = \frac{1}{B_t(s)}\left\{\frac{\partial B^T}{\partial Z}\eta_Z + \frac{\partial B}{\partial t} + \frac{1}{2}trace(\sigma_Z \sigma_Z^T \frac{\partial^2 B}{\partial Z^2})\right\},$$

$$\sigma_B(t, s) = \frac{1}{B_t(s)}\sigma_Z^T \frac{\partial B}{\partial Z}.$$

Proof This theorem follows by a direct application of Itô's formula (Theorem A.3.13) to the given setting.

Lemma 9.4.2 *Suppose the economic model does not allow arbitrage opportunities. Then there exist functions $(\lambda_i(Z_t, t))_{i=1,...,d}$ which satisfy*

$$\lambda^T \sigma_B(t, s) = \eta_B(t, s) - r_t, \text{ with } 0 \leq s < t.$$

Proof [Vas77] and [CIR85].

Remark 9.4.3 • In the following we assume that the functions λ are independent of the occurrence of the insured event.
• We also assume that they satisfy Novikov's condition:

$$E^{P_1}\left[\exp(\frac{1}{2}\int_0^T \lambda^T \lambda du)\right] < \infty. \tag{9.3}$$

Theorem 9.4.4 *Suppose the economic model for the interest rates does not allow arbitrage and Novikov's condition holds. Then the present value of a zero coupon bond with maturity s at time t is*

$$B_t(s) = E\left[\exp(-\int_t^s r_u du) \times \exp(-\int_t^s \lambda^T dW - \frac{1}{2}\int_t^s \lambda^T \lambda\, du) \mid \mathcal{G}_t\right].$$

Remark 9.4.5 After proving the above theorem, we can define the equivalent martingale measure by

$$\xi_t = \frac{dQ}{dP} = \exp(-\int_t^s \lambda^T dW - \frac{1}{2}\int_t^s \lambda^T \lambda\, du)$$

and it satisfies:

1. $\xi_t \geq 0$,
2. $E^{P_1}[\xi_t] = 1$,
3. $Q_1(D) = E[\chi_D \xi_T]$ for all $D \in \mathcal{G}$.

Proof We define

$$A(u) = -\int_t^u \left(r_v + \frac{1}{2}\lambda^T \lambda\right) dv - \int_t^u \lambda^T dW,$$
$$Y(u) = B_u(s)\exp(A(u)).$$

Next we consider $Y(u) := h(B_u(s), \exp(A(u)))$ and calculate the drift dY, using the fact that

$$dA = -(r + \frac{1}{2}\lambda^T \lambda)du + \lambda^T dW$$

and

$$dB = \eta_B\, B du + B\, \sigma_B^T dW$$

hold. It turns out that this term is equal to zero. Therefore, $Y(u)$ is a martingale and we get:

$$E[Y(s) \mid \mathcal{F}_t] = Y(t).$$

Now, $Y(t) = B_t(s)$ and

$$Y(s) = E\left[\exp(-\int_t^s r_u)\, du \times \exp(-\int_t^s \lambda^T dW - \frac{1}{2}\int_t^s \lambda^T \lambda\, du) \mid \mathcal{G}_t\right],$$

which implies the result.

Using the previous theorem we can calculate the price of a zero coupon bond.

Lemma 9.4.6
$$B_t(s) = E^{Q_1}\left[\exp\left(-\int_t^s r_u du\right) \mid \mathcal{G}_t\right].$$

Proof We set

$$\xi_{t,s} = \exp\left(-\int_t^s \lambda\, dW - \frac{1}{2}\int_t^s \lambda^T \lambda\, du\right).$$

One can calculate the expectation of this, and it turns out to be $E^{P_1}\left[\xi_{t,T}\right] = 1$ for $t \le T$. Furthermore, we know that

$$E^{Q_1}\left[\exp\left(-\int_t^s r_u du\right) \mid \mathcal{G}_t\right]$$
$$= E\left[\exp\left(-\int_t^s r_u du\right)\xi_{0,T} \mid \mathcal{G}_t\right] \Big/ E\left[\xi_{0,T} \mid \mathcal{G}_t\right].$$

The individual terms of this can be calculated by

$$E^{P_1}\left[\xi_{0,T} \mid \mathcal{G}_t\right] = \xi_{0,t} \times E^{P_1}\left[\xi_{t,T} \mid \mathcal{G}_t\right]$$
$$= \xi_{0,t}$$

and

$$E^{P_1}\left[\exp\left(-\int_t^s r_u du\right)\xi_{0,T} \mid \mathcal{G}_t\right]$$
$$= E^{P_1}\left[\exp\left(-\int_t^s r_u Du\right)\xi_{0,s} E^{P_1}[\xi_{s,T}|\mathcal{G}_s] \mid \mathcal{G}_t\right]$$
$$= E^{P_1}\left[\exp\left(-\int_t^s r_u du\right)\xi_{0,s} \mid \mathcal{G}_t\right]$$
$$= \xi_{0,t} \times E^{P_1}\left[\exp\left(-\int_t^s r_u du\right)\xi_{t,s} \mid \mathcal{G}_t\right].$$

Finally, we get

$$B_t(s) = \frac{\xi_{0,t} \times E^{P_1}\left[\exp\left(-\int_t^s r_u du\right)\xi_{t,s} \mid \mathcal{G}_t\right]}{\xi_{0,t}}$$
$$= E^{Q_1}\left[\exp\left(-\int_t^s r_u du\right) \mid \mathcal{G}_t\right].$$

Thus, we have seen that it is possible to choose $\frac{dP}{dQ}$ as shown above. Now, this equation and an application of Girsanov's theorem yield the following representations with respect to Q:

$$Z_t = Z_0 + \int_0^t [\eta_Z \sigma_Z \lambda] \, du + \int_0^t \sigma_Z^T d\widehat{W},$$

$$B_t(s) = B_0(s) + \int_0^t r_u \, B_u(s) du + \int_0^t B_u(s) \sigma_B^T d\widehat{W},$$

where $\widehat{W}_s$ is a standard Brownian motion with respect to Q.

9.5 Thiele's Differential Equation

We use the interest rate model from the previous section. Furthermore we assume that the insurance model is regular (Definition 4.5.6). The mathematical reserve is defined by

$$V_t^g = \int_t^T \Pi_t^g(u) du,$$

where

$$\Pi_t^g(u) = B_t(u) \times P_t^g(u),$$

$$P_t^g(u) = \sum_j p_{gj}(t, u) \left\{ a_j(u) + \sum_{k \neq j} \mu_{jk}(u) a_{jk}(u) \right\}.$$

To derive Thiele's differential equation, we calculate by the chain rule the partial derivative of the mathematical reserve with respect to time:

$$\frac{\partial}{\partial t} V_t^g = \frac{\partial}{\partial t} \int_t^T \Pi_t^g(u) du,$$

$$\int_t^T \frac{\partial}{\partial t} \Pi_t^g(u) du = \frac{\partial V_t^g}{\partial t} + a_g(t) + \sum_{k \neq g} \mu_{gk}(t) a_{gk}(t).$$

Moreover

$$B_t(u) = \frac{\Pi_t^g(u)}{P_t^g}.$$

Now Kolmogorov's theorem implies

$$\frac{\partial}{\partial t} p_{gj}(t, u) = \sum_{k \neq g} \mu_{gk}(t) \left\{ p_{gj}(t, u) - p_{kj}(t, u) \right\}$$

and we get

$$\frac{\partial}{\partial t} B_t(u) = \frac{\partial}{\partial t} \frac{\Pi_t^g(u)}{P_t^g}$$

$$= \frac{1}{P_t^g} \left(\sum_{k \neq g} \mu_{gk}(t) \left\{ \Pi^k(u) - \Pi^g(u) \right\} + \frac{\partial}{\partial t} \Pi_t^g \right).$$

But we also know that Π_t^g is a function of Z. Thus the drift term of B_t can, by Itô's formula, be calculated as follows:

$$\frac{1}{P_t^g} \left(\frac{\partial \Pi^g(u)^T}{\partial Z} (\eta_Z - \sigma_Z \lambda) + \frac{1}{2} \mathrm{trace}(\sigma_Z \sigma_Z^T \frac{\partial^2 \Pi_t^g(u)}{\partial Z^2}) \right.$$

$$\left. + \sum_{k \neq g} \mu_{gk}(t) \left\{ \Pi^k(u) - \Pi^g(u) \right\} + \frac{\partial}{\partial Z} \Pi_t^g \right).$$

Using the stochastic differential equation which defines B, we know that this term has also the representation:

$$r_t B_t(u) = \frac{1}{P_t^g} \Pi_t^g(u) r_t.$$

Finally, the uniqueness of the drift term implies the following partial differential equation for Π^g:

$$\frac{\partial \Pi^g(u)^T}{\partial Z} (\eta_Z - \sigma_Z \lambda) + \frac{1}{2} \mathrm{trace}(\sigma_Z \sigma_Z^T \frac{\partial^2 \Pi_t^g(u)}{\partial Z^2})$$

$$+ \sum_{k \neq g} \mu_{gk}(t) \left\{ \Pi^k(u) - \Pi^g(u) \right\} + \frac{\partial}{\partial Z} \Pi_t^g - \Pi^g(u) r_u = 0.$$

This equation yields Thiele's differential equation for the given setting.

Theorem 9.5.1 (Thiele's differential equation) *Let a regular insurance model and the interest rate process defined in Sect. 9.4 be given. Then the following partial differential equation holds:*

$$\frac{\partial V_t^g}{\partial t} = r_t V_t - a_g(t) - \sum_{k \neq g} \mu_{gk}(t) \left\{ a_{gk}(t) + V_t^k - V_t^g \right\}$$

$$- \left[\frac{\partial V_t^g}{\partial Z} (\eta_Z - \sigma_Z \lambda) + \frac{1}{2} \mathrm{trace}(\sigma_Z \sigma_Z^T \frac{\partial^2 \Pi_t^g(u)}{\partial Z^2}) \right].$$

Remark 9.5.2 The differential equation in the theorem above consists of three parts:

1. classical part: $r_t \, V_t^g - a_g(t) - \sum_{k \neq g} \mu_{gk}(t) \{a_{gk}(t) + V_t^k - V_t^g\}$
2. component corresponding to the stochastic interest rate:

$$-\left[\frac{\partial V_t^g}{\partial Z}(\eta_Z - \sigma_Z \, \lambda) + \frac{1}{2}\mathrm{trace}(\sigma_Z \sigma_Z^T \frac{\partial^2 \Pi_t^g(u)}{\partial Z^2})\right]$$

3. The additional term $-\sigma_Z \lambda$ appears in the formula, since we had to use the equivalent martingale measure instead of the original measure. If one would calculate everything with respect to the original measure, then this term would disappear.

Example 9.5.3 We look again at the interest rate model of Vasiček [Vas77] (Sect. 9.2). Therein η and σ are given by the functions

$$\eta(r_t, t) = \alpha\,(\rho - r_t),$$
$$\sigma(r_t, t) = \sigma.$$

Thus we have the stochastic differential equation

$$dr_t = \alpha(\rho - r_t)dt + \sigma\,dW_t.$$

If we assume that $\lambda = \lambda(r_t, t)$, then we get

$$B_0(t) = E^Q[\exp(-\int_0^t r_s ds)] = G_t \, \exp(-H_t\, r_t),$$

where

$$H_t = \frac{1 - \exp(-\alpha t)}{\alpha},$$
$$G_t = \exp((\rho - \frac{\lambda \sigma}{\alpha} - \frac{1}{2}(\frac{\sigma}{\alpha})^2)(H_t - t) - \frac{1}{\alpha}(\frac{\sigma H_t}{2})^2)$$

(compare with [Vas77]).

Now these formulas together with Thiele's differential equation yield for an endowment policy the equation

$$\frac{\partial \Pi}{\partial t} = (\mu_{x+t} + r_t)\Pi_t + \bar{p}_t - c_t\, \mu_{x+t} - \left[\frac{\partial \Pi}{\partial r}(\alpha(\rho - r) - \sigma\lambda) + \frac{1}{2}\sigma^2\frac{\partial^2 \Pi}{\partial r^2}\right].$$

Chapter 10
Technical Analysis

In this chapter we discuss the technical analysis of life insurance policies. This is the actual technical analysis which has to be done at the end of each fiscal year. It is concerned with the question, if the underlying loaded assumptions resemble the reality or if these have to be adapted. This analysis is used for yearly inspections of the current portfolio.

At the end of this chapter we will look at profit testing and the calculation of the present value of the future profits. These techniques are required for the calculation of the profit potential of the insurance products.

10.1 Classical Technical Analysis

In the technical analysis of an insurance policy one compares, for a given period, the incurred losses with the the funds available to cover these.

First, one needs methods to calculate the effective incurred losses and the premiums for the policies. The latter will be split into a risk premium and a savings premium.

In order to begin the technical analysis one has to define the normal subsequent state for each state $j \in S$. The definition of the subsequent state depends on the model. This state describes the transitions without incurred losses. In the setting of a classical life insurance with state space $S = \{*, \dagger\}$ the subsequent state of $*$ is usually defined to be $*$.

Definition 10.1.1 (*Normal subsequent state*) The normal subsequent state is defined by a function

$$\phi : S \to S, i \mapsto \phi(i),$$

which assigns to each state i, a state which is subsequent to it and for which, according to the policy setup, no payout to the insured is due.

M. Koller, *Stochastic Models in Life Insurance*, EAA Series,
DOI: 10.1007/978-3-642-28439-7_10, © Springer-Verlag Berlin Heidelberg 2012

One can derive the following quantities based on the normal subsequent state.

Definition 10.1.2 (*Savings premium*) The savings premium for state i during the time interval $]t, t + 1]$ is denoted by $\Pi_i^{(s)}$. It can be calculated by the following formula:

$$\Pi_i^{(s)}(t) = v_t^i \, V_{\phi(i)}(t + 1) - V_i(t).$$

The savings premium is the amount of money, which one has to add to the mathematical reserve at time t in order to have the necessary mathematical reserve in the subsequent state $\phi(i)$ at time $t + 1$.

Definition 10.1.3 (*Technical pension dept*) The regular cash flow or the technical pension dept is

$$\Pi_i^{(tr)}(t) = a_i^{Pre}(t) + v_t^i \, a_{i\phi(i)}^{Post}(t).$$

For a policy with premiums the technical pension dept $\Pi_i^{(tr)}(t)$ and the premiums coincide.

Definition 10.1.4 (*Value at risk and risk premium*) Let $i \in S$ and $j \neq \phi(i)$. Then the value at risk $R_{ij}(t)$ is given by

$$R_{ij}(t) = V_j(t + 1) + a_{ij}^{Post}(t) - (V_{\phi(i)}(t + 1) + a_{i\phi(i)}^{Post}(t)).$$

It corresponds to the loss, which incurs for a transition $i \rightsquigarrow j$. The risk premium corresponding to the transition $i \rightsquigarrow j$ is given by

$$\Pi_{ij}^{(r)}(t) = p_{ij}(t) \, v_t^i \, R_{ij}(t).$$

Furthermore, $\Pi_i^{(r)}(t) = \sum_{j \neq \phi(i)} \Pi_{ij}^{(r)}(t)$ denotes the total risk premium.

Based on these quantities one can perform a technical analysis by comparing the loss, which effectively incurred in the year, and the risk premium for the corresponding transition. Usually one also considers the loss quotient $\frac{\text{loss}(i \rightsquigarrow j)}{\Pi_{ij}^{(r)}}$.

Example 10.1.5 If for a life insurance the mortality rate is 15% above the effectively observed mortality, then the loss quotient is roughly 85% ($= 1/1.15$).

We have seen that one can use the risk premium and the losses to check if the underlying best estimate assumptions fit to the current situation.

The following theorem relates the two kinds of premiums.

Theorem 10.1.6 *The regular cash flow is related to the savings premium and the risk premium by the following equation:*

$$-\Pi_i^{(tr)}(t) = \Pi_i^{(r)}(t) + \Pi_i^{(s)}(t).$$

This decomposition of the premiums is called technical decomposition.

Proof The statement is proved by a direct calculation:

$$
\Pi_i^{(r)}(t) + \Pi_i^{(s)}(t)
$$
$$
= \sum_{j \in S} v_t \, p_{ij}(t) \left(V_j(t+1) + a_{ij}^{Post}(t) - V_{\phi(i)}(t+1) - a_{i\phi(i)}^{Post}(t) \right)
$$
$$
\quad + (v_t V_{\phi(i)}(t+1) - V_i(t))
$$
$$
= -v_t \left(V_{\phi(i)}(t+1) + a_{i\phi(i)}^{Post}(t) \right) \times p_{i\phi(i)}(t) - v_t \left(V_{\phi(i)}(t+1) + a_{i\phi(i)}^{Post}(t) \right)
$$
$$
\quad \times (1 - p_{i\phi(i)}(t)) + v_t V_{\phi(i)}(t+1) - a_i^{Pre}(t)
$$
$$
= -\Pi_i^{(tr)}(t),
$$

where we used Thiele's difference equations.

Remark 10.1.7 Theorem 10.1.6 states that the premium can be decomposed into a savings premium and a risk premium. This holds for net premiums without any administration charge. The gross premium is composed of three parts: savings premium, risk premium, cost premiums.

10.2 Profit Testing

The analysis of an insurance product with respect to a given period is called profit testing. For this one tries to calculate for a period $[t, t+1[$ the return of a policy depending on various parameters.

The return, which is generated by selling a policy, mainly depends on the following effects:

- The major part of the return, which is induced by the insurance contract, is usually due to the return on investment. For example a mathematical reserve of 100,000 USD and a return on investment of 5.1% (technical interest rate: 3.5%) yields 100 USD per year, with an payed out surplus of 1.5%.
- Besides the interest surplus also the possible risk surplus is relevant. Also in this case it is possible, that the insured does not receive the total risk surplus.
- Finally one also has to consider the effect of a redemption of the insurance. This yields a return, since in the case of redemption the insured does not get back the total mathematical reserve. He has to accept a reduction. For example suppose that 5% of all insurances are redeemed each year and that the redemption value is 97% of the mathematical reserve. Then the expected return of a redemption is 0.15% of the mathematical reserve.
- Furthermore, return might be generated by the difference of the charged administration costs and the effective administration costs.

One has to note, that not all of the above returns have to be positive, i.e. some might actually reduce the total return. Consider for example an insurance company which guarantees an interest surplus which is much higher than the effectively realised return. Then clearly the return on investment is negative.

The aim of profit testing is to derive a profile of the returns with respect to time. Furthermore the effects of various parameters on the return profile are determined. Thus a model is used to calculate the expected yearly returns. Then one looks for example at the effect of an 10% increase of the mortality.

Certainly we can not provide in this book a general model for the return of an insurance company, since for this specific details of the each company have to be taken into account. Nevertheless, note the following list of effects which are usually considered:

Profit and loss based on

- return on investment,
- risk,
- redemption,
- acquisition costs,
- administration costs,
- taxes.

10.3 Embedded Value

In this section the concept of the embedded value will be introduced. This quantity is used to measure the value of an insurance company, of a division of the company or of a product. Therefore the embedded value is of major interest to investors and to the management of the company. Besides measuring the value of the company this quantity can also be used to determine the salaries of the management or of the sales staff. In fact there are more and more companies which link the salaries of the management to the embedded value.

The concept of the embedded value is very popular in Anglo-Saxon countries. Also in continental Europe this method is becoming increasingly important.

The embedded value is composed of the present value of the future profits of all insurance policies and of the net asset value of the company. The net asset value is the value of the insurance company, which are not bound to outstanding debts, especially they are not part of the mathematical reserves. They originate form parts of the profits which were not distributed to the shareholders and the paid in share capital. Usually one can determine the net asset value with the help of the accounts of the company. Besides the net asset value the second component of the embedded value is the so called present value of future profits (pvfp). This quantity has to be calculated with the help of models. Here the Markov model will be used to derive a model for the return process.

At this point one should note, that the model for the calculation of the pvfp varies for the various types of policies. Thus its calculation strongly depends on the policy. Moreover the complexity of the chosen model might vary strongly with the aims of the calculation, it also depends on regional characteristics and the specific setting of the policy and the insurance company. Therefore the difficulty in calculating the pvfp is not caused directly by the model, but it is due to the diversity of possible models and the wide range of insurance products.

We start with a collection of factors which have an influence on the return of an insurance company. One might have to find a model which incorporates one or more of the following:

- return on investment for bonds and shares,
- return based on risks,
- return based on costs,
- reinsurance,
- redemption,
- surplus.

To incorporate these effects into a Markov model the following terms have to be defined accordingly:

- state space,
- discount functions,
- policy functions.

Given these functions one can easily apply the theory which we developed before. For example one can calculate the distribution or the moments of the pvfp. These quantities capture the temporal variability of the pvfp. Moreover, an analysis could also consider a whole portfolio, and thus it could determine the pvfp and its characteristics for all policies hold by an insurance company.

Now we will give precise definitions of the components of the model. Afterwards, we have a look at some examples.

10.3.1 State Space

To calculate the pvfp one has to consider a state space which is suitable for the problem at hand. Thus, first of all the state space depends on the underlying insurance policy. It also depends on the modelling techniques which we want to apply, i.e. do we use a deterministic or a stochastic model for the return on investment. In the latter case the state space would be larger. For example, consider a life insurance with state space $\{*, †\}$. Now suppose we want to include the possibility of redemption and a stochastic interest rate model with Markovian interest intensities (state space S_W). Then, for the calculation of the pvfp, we have to use the state space $\{*, †, ‡\} \times S_W$, where $‡$ denotes the state of a lapsed policy.

To ensure that the model (state space S_1) for the calculation of the premiums and the model (state space S_2) for the calculation of the embedded value are consistent, we have to assume that a mapping

$$\Phi : S_2 \to S_1, s_2 \mapsto s_1$$

exists. In order to satisfy this assumption it might be necessary to enlarge the state space S_1 with states which are actually not required for the calculation of the premiums.

In the above example the state $\ddagger$ has to be added to the state space S_1, since redemption was not explicitly considered in the calculation of the premiums. Then one can define the mapping $\Phi : S_2 = S_1 \times S_W \to S_1$ by $(i, j) \in S_1 \times S_W \mapsto i$. Thus S_2 can be visualised as a decomposition of the states of S_1, where Φ determines the relation of the states within the models.

10.3.2 Discount Function

In this section we want to define the discount function. First of all one should note that there are several approaches to this problem. The classical approach tries to determine the value of the insurance company directly. For this one imagines an investor who wants to get a return on his investment, and the return should be reasonably related to the risks involved. Thus one defines the risk discount rate, which is basically the return of an investment into something (e.g. shares) with a similar risk profile. In fact, the return of shares of the corresponding market is usually a good reference point. Note that this incorporates the fact that the risk discount rate depends on the development of the economy.

The risk discount rate is just one of several possibilities to calculate the discount. For example, if one wants to calculate the expected value of an insurance company and, at the same time, derive its risk profile, then one would have to use a stochastic interest rate model. Here the models which were introduced in the previous chapters are applicable.

In any case, one should note that the pvfp strongly depends on the choice of the discount function.

10.3.3 Definition of the Policy Functions

Next, given the state space and the discount function, the policy functions will be defined. For this one calculates the profit or loss during a given time interval for every transition between states. Depending on the required precision and on the purpose of the whole model one decides which effects should be incorporated into the model. Usually one focuses on those effects, which have a high impact on the return. Hence, for many policies the focus is on the modelling of the return of investment in shares and bonds and the integration of the risk of redemption.

Now, the question is how to calculate the profit and loss for a given state transition of the policy. For this one has to analyse the cash flows, which occur within a year. It is helpful to assume that the insurance company has the mathematical reserve $V_{\Phi(i)}(t)$ at the beginning of the year and also receives at that time the corresponding premiums. At the end of the year the mathematical reserve $V_{\Phi(j)}(t+1)$ has to be stored and various expenses like administration costs and insurance benefits have to be financed. The following table illustrates this situation:

Time	t	$t+1$	Comment
t	$V_{\Phi(i)}(t) + P$		Income: reserve and premium
t	$-a_{\Phi(i)}^{Pre}(t)$		Expenses: pension
t	costs		Expenses: costs
t	$\pm\ldots$		Other
$t+1$		$-V_{\Phi(j)}(t+1)$	Expenses: reserve
$t+1$		$-a_{\Phi(i)\Phi(j)}^{Post}(t)$	Expenses: capital
$t+1$		$-$ surplus	Expenses: surplus
$t+1$		$\pm\ldots$	Other

10.3.4 Examples

Previously we have discussed the general setup for the calculation of the pvfp. Now we will consider concrete examples.

Example 10.3.1 The first example is an endowment policy, which was bought for a single premium of 100,000 USD. We suppose the insured is 40 years old and the policy matures in 10 years. Since the policy was bought for a single premium, we suppose there is no reduction in the case of redemption, i.e., the insurance company does not make any additional profit if the policy is redeemed. Furthermore we assume:

- The yearly administration costs are fixed within the policy to 0.6% of the mathematical reserve plus an additional 100 USD.
- The costs for the contract are 5% of the single premium plus an additional 200 USD.
- The effective yearly administration costs are 210 USD, and by inflation they increase by 3.25% per year.
- The effective costs for the contract (commission) are 7% of the single premium.
- We assume a technical interest rate of 4% and a return on investment for shares of 5%. The insured gets 95% of the interest surplus, the remaining 5% are profit for the insurer.
- The risk spread is 10%, and we assume that 90% of this are credited to the insured.

Table 10.1 Technical values of an endowment policy

Age	q_x	$_1p_x$	Death benefit	Mathematical reserve	Cost premium	Risk premium
40	0.0011	0.9989	130064	100000	5, 950.00	35.04
41	0.0012	0.9988	130064	97,12	736.87	34.46
42	0.0013	0.9987	130064	100927	755.56	33.39
43	0.0015	0.9985	130064	104149	774.89	31.92
44	0.0016	0.9984	130064	107480	794.88	29.81
45	0.0018	0.9982	130064	110927	815.56	27.10
46	0.0020	0.9980	130064	114492	836.95	23.08
47	0.0022	0.9978	130064	118183	859.10	17.44
48	0.0025	0.9975	130064	122004	882.02	9.82
49	0.0027	0.9973	130064	125962	905.77	––

Table 10.2 Profit/loss profile of an endowment policy with a single premium

Age	Return (%)	s_x	$_tp_x$	p/l costs	p/l risk	p/l investment	p/l total
40	5	0.2000	1.0000	−1260.00	3.50	46.11	−1, 210.39
41	5	0.1400	0.7989	371.84	2.75	38.76	413.36
42	5	0.0800	0.6860	294.48	2.29	34.35	331.12
43	5	0.0800	0.6302	257.22	2.01	32.57	291.80
44	5	0.0800	0.5789	221.49	1.73	30.87	254.09
45	5	0.0800	0.5316	187.17	1.44	29.26	217.88
46	5	0.0800	0.4882	154.14	1.13	27.74	183.00
47	5	0.0800	0.4481	122.28	0.78	26.28	149.35
48	5	0.0800	0.4113	91.51	0.40	24.90	116.82
49	5	0.0800	0.3773	61.73	––	23.59	85.33
Total				122.82	13.54	245.51	**381.87**

- We assume a risk discount rate of 7% and denote the yearly redemption probability by s_x.

Based on these assumptions 130,064 USD is the death benefit and endowment, respectively, for the given policy. Further values of this calculation are shown in Table 10.1, the resulting profits and losses are listed in Table 10.2. An analysis of these values shows a correlation between the return and the age of the policy. Especially the loss in the first year stands out, it is caused by the commission. Figure 10.1 illustrates the same values in form of a graph.

For the previous example the present value of the future profits is 381.87 USD, of theses 122.82 USD are due to profits with respect to administration, 13.54 USD are due to the risks and 245.51 US are due to the return on investment in shares. The irr (internal rate of return) for this policy is about 8.7%. In a next step one can try to analyse the sensitivity of the pvfp with respect to various parameters. Most of the profit in this example is due to the administration of the policy. If we are

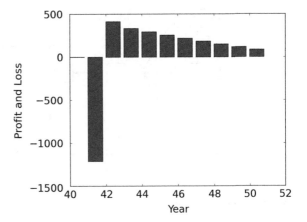

Fig. 10.1 Profit/loss profile of a endowment policy

interested in the relation between the present value of the future profits and s_x we just multiply s_x by a factor γ. Then we note, that the profit would increases to 773.44 USD, if one could reduce the redemptions by 20% ($\gamma = 0.8$). On the other hand a 20% increase of the redemption probability would yield a profit of only 28.05 USD. These values illustrate the usefulness of the calculations, since they provide a tool for the management of the company to assess the consequences of certain decisions.

After the calculation of the pvfp for the endowment policy we will now look at a more complicated example. We consider a disability pension as introduced in Sect. 6.6. This type of insurance is particular difficult to handle, since the probability of becoming disabled depends on the state of the economy and it varies over time. We will develop a model for this type of policy in the following example.

Example 10.3.2 To analyse the influence of various effects one has to generalise the disability pension model. The state $*$ will be decomposed into a set of states $*_1$ to $*_m$. For a state $*_k$ the probability of becoming disabled $p_{*_k \diamond_1}(x)$, denoted by i_x, is much larger than i_x^{Ref}, which is the assumed probability of becoming disabled (used for the calculation of the premiums). Additionally we assume that the transitions between the states $*_k$ are given by a homogeneous Markov chain. The disability pension model, which we will use for the calculation of the premium, is the same as in the example in Sect. 6.6. We work with $m = 6$ states of disability. For the calculation of the pvfp we have to define the corresponding model. We only look at those profits and losses which are due to a change of the probability of becoming disabled. The corresponding state space is depicted in Fig. 10.2. Further we make the following assumptions:

- The disability pension is fixed to 10,000 USD and it has to be payed yearly in advance.
- Technical interest rate 4%. Risk discount rate 8%.
- We consider the case $m = 6$ and assume that $i_x^k = \gamma_k \times i_x^{\mathrm{Ref}}$, where $\gamma = \{1.5, 1.3, 1.1, 0.9, 0.7, 0.5\}$. This means for example that the effective proba-

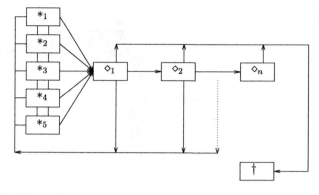

Fig. 10.2 Disability model

bility of becoming disabled in state $*_4$ is 90% of the corresponding probability used for the calculation of the premiums.

- The transition probabilities $p_{*_i,*_k}$ are given by

$$P = \begin{pmatrix} 0.70 & 0.20 & 0.10 & -- & -- & -- \\ 0.10 & 0.70 & 0.10 & 0.10 & -- & -- \\ 0.10 & 0.10 & 0.60 & 0.10 & 0.10 & -- \\ -- & 0.10 & 0.10 & 0.60 & 0.10 & 0.10 \\ -- & -- & 0.10 & 0.10 & 0.70 & 0.10 \\ -- & -- & -- & 0.10 & 0.20 & 0.70 \end{pmatrix}.$$

- In case of a reactivation the new state is one of the six active states. The state is chosen with probability 1/6.
- We consider a 30 year old person and suppose that 65 is the age of maturity of the policy.

It is useful to look at an example to understand the calculation of the pvfp. The mathematical reserve for a 55 year old active person is 9,607 USD. Then for a 56 year old active person it is 8,352 USD, but in case of disability it is 57,353 USD. Thus, if the man is 55 years old he might make the following loss or profit (at the end of the year):

From	To	Profit/loss
$*_k$	$*_l$	1,702 USD
$*_k$	†	10,376 USD
$*_k$	$\diamond_1$	−49,183 USD

At the end of the year, the loss which incurred by becoming disabled can be calculated as follows:

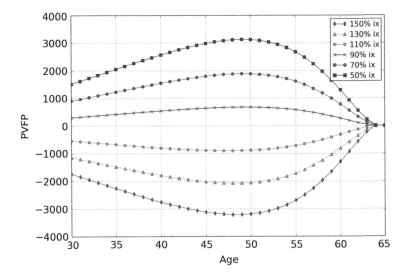

Fig. 10.3 Profit/loss profile of a disability insurance

	On January 1st	On December 31st
Reserve begin of year	9,607 USD	
Reserve for loss	−55,147 USD	−57,353 USD
Sum begin of year	−45,540 USD	
Sum end of year		**− 49,183 USD**

Using this scheme one can calculate the profit or loss for every transition, and Thiele's difference equations yield for $x = 30$ the following values:

State	Level	pvfp
*1	$150\% \times i_x$	−1732.50 USD
*2	$130\% \times i_x$	−1138.26 USD
*3	$110\% \times i_x$	−538.06 USD
*4	$90\% \times i_x$	295.12 USD
*5	$70\% \times i_x$	902.39 USD
*6	$50\% \times i_x$	1515.75 USD

This calculations show that the present value of future profits strongly depends on the probability of becoming disabled. Suppose a loss ratio of 90%, then the pvfp is about 300 USD, which is about 3% of one pension payment. If the loss ratio increases by 20%, the present value of the loss is already about 5% of one pension payment.

This analysis can be used to determine the necessary spread in the premium rate. The dependence of the pvfp on the level of the probability of becoming disabled is shown in Fig. 10.3.

For further informations related to the embedded value we refer to [CFO08].

Chapter 11
Abstract Valuation

11.1 Framework: Valuation Portfolios

Definition 11.1.1 (*Stochastic Cash Flows*) A *Stochastic Cash Flow* is a sequence $x = (x_k)_{k \in \mathbb{N}} \in L^2(\Omega, \mathcal{A}, P)^{\mathbb{N}}$, which is $\mathbb{F} = (\mathcal{F}_t)_{t \geq 0}$ adapted.

Definition 11.1.2 (*Regular Stochastic Cash Flows*) A *Regular Stochastic Cash Flow* x with respect to $(\alpha_k)_{k \in \mathbb{N}}$, with $\alpha_k > 0 \forall k$ is a stochastic cash flow such that

$$Y := \sum_{k \in \mathbb{N}} \alpha_k X_k \in L^2(\Omega, \mathcal{A}, P).$$

We denote the vector space of all regular cash flows by $\mathcal{X}$.

Remark 11.1.3 1. We note that for all $n \in \mathbb{N}_0$ the image of

$$\psi : L^2(\Omega, \mathcal{A}, P)^n \to \mathcal{X}, (x_k)_{k=0,\dots n} \mapsto (x_0, x_1, \dots, x_n, 0, 0 \dots)$$

is a sub-space of $\mathcal{X}$.

2. $\mathcal{X}$ has been defined this way in order to capture cash flow streams where the sum of the cash flows is infinite with a finite present value. In this set up α_k can be interpreted as a majorant of the price of the payment 1 at time k.

Theorem 11.1.4 *1. For $x, y \in \mathcal{X}$, we define the scalar product as follows:*

$$< x, y > = \sum_{k \in \mathbb{N}} < \alpha_k x_k, \alpha_k y_k >$$
$$= E[\sum_{k \in \mathbb{N}} \alpha_k^2 x_k y_k],$$

and remark that the scalar product exists as a consequence of the Cauchy-Schwartz inequality .

M. Koller, *Stochastic Models in Life Insurance*, EAA Series,   145
DOI: 10.1007/978-3-642-28439-7_11, © Springer-Verlag Berlin Heidelberg 2012

2. $\mathcal{X}$ *equipped with the above defined scalar product is a Hilbert space with norm* $||x|| = \sqrt{<x,x>}$.

Proof We leave the proof of this proposition to the reader.

In a next step we introduce the concept of a positive valuation functional and we closely follow [Büh95].

Definition 11.1.5 (*Positivity*)

1. $x = (x_k)_{k\in\mathbb{N}} \in \mathcal{X}$ is called *positive* if $x_k > 0$ P-a.e. for all $k \in \mathbb{N}$. In this case we write $x \geq 0$.
2. $x = (x_k)_{k\in\mathbb{N}} \in \mathcal{X}$ is called *strictly positive* if $x_k > 0$ P-a.e. for all $k \in \mathbb{N}$ and there exists a $k \in \mathbb{N}$, such that $x_k > 0$ with a positive probability. In this case we write $x > 0$.

Definition 11.1.6 (*Positive functionals*) $Q : \mathcal{X} \to \mathbb{R}$ is called a *positive, continuous and linear functional* if the following hold true:

1. If $x > 0$, we have $Q[x] > 0$.
2. If $x = \lim_{n\to\infty} x_n$, for $x_n \in \mathcal{X}$ we have $Q[x] = \lim_{n\to\infty} Q[x_n]$.
3. For $x, y \in \mathcal{X}$ and $\alpha, \beta \in \mathbb{R}$ we have $Q[\alpha x + \beta y] = \alpha Q[x] + \beta Q[y]$.

Theorem 11.1.7 (Riesz representation theorem) *For Q a positive, linear functional as defined before, there exists* $\phi \in \mathcal{X}$, *such that*

$$Q[y] = <\phi, y> \forall y \in \mathcal{X}.$$

Proof This is a direct consequence of Riesz representation theorem of continuous linear functionals of Hilbert spaces.

Definition 11.1.8 (*Deflator*) The $\phi \in \mathcal{X}$ generating $Q[\bullet]$ is called *deflator*.

Theorem 11.1.9 *For a positive functional* $Q : \mathcal{X} \to \mathbb{R}$, *with deflator* $\psi \in \mathcal{X}$ *we have the following:*

1. $\phi_k > 0$ *for all* $k \in \mathbb{N}$.
2. ϕ *is unique.*

Proof 1. Assume $\phi_k = 0$. In this case we have $Q[(\delta_{kn})_{n\in\mathbb{N}}] = 0$ which is a contradiction.
2. Assume $Q[y] = <\phi, y> = <\phi^*, y>$ for all $y \in \mathcal{X}$. In this case we have $<\phi - \phi^*, y> = 0$, in particular for $y = \phi - \phi^*$. Hence we have $||\phi - \phi^*|| = 0$.

Definition 11.1.10 (*Projections*) For $k \in \mathbb{N}$ we define the following projections:

1. $p_k : \mathcal{X} \to L^2(\Omega, \mathcal{A}, P), x = (x_n)_{n\in\mathbb{N}} \mapsto (\delta_{kn} x_n)_{n\in\mathbb{N}}$, the projection on the k-th coordinate.
2. $p_k^+ : \mathcal{X} \to L^2(\Omega, \mathcal{A}, P), x = (x_n)_{n\in\mathbb{N}} \mapsto (\chi_{k\leq n} x_n)_{n\in\mathbb{N}}$, the projection starting on the k-th coordinate.

Definition 11.1.11 (*Valuation at time t, pricing functionals*) For $t \in \mathbb{N}$ we define the valuation of $x \in \mathcal{X}$ at time t by

$$Q_t[x] = Q[x|\mathcal{F}_t] = \frac{1}{\phi_t} E[\sum_{k=0}^{\infty} \phi_k \, x_k|\mathcal{F}_t]$$

In the same sense as for mathematical reserves we define the value of the future cash flows at time t by

$$Q_t^+[x] = Q[p_t^+(x)].$$

The operators Q_t and Q_t^+ are called pricing functionals.

Definition 11.1.12 (*Zero Coupon Bonds*) The Zero Coupon Bond $\mathcal{Z}_{(k)} = (\delta_{kn})_{n\in\mathbb{N}}$ is an element of $\mathcal{X}$. We remark that

$$\pi_0\left(\mathcal{Z}_{(t)}\right) = Q[\mathcal{Z}_{(t)}] = E[\phi_t].$$

Theorem 11.1.13 *The cash flow* $x = (x_k)_{k\in\mathbb{N}}$ *in the discrete Markov model (cf. Proposition 4.7.3) on a finite time interval* $T \subset \mathbb{N}$ *is given by:*

$$x_k = \sum_{(i,j)\in S^2} \Delta N_{ij}(k-1)a_{ij}^{Post}(k-1) + \sum_{i\in S} I_i(k)a_i^{Pre}(k),$$

where we assume that $\Delta N_{ij}(-1) = 0$.

Proof The form of the cash flow follows from the calculations of the earlier chapters. It remains to show that $(x_k)_{k\in\mathbb{N}}$ is in L^2. This is however easy, since the benefit functions and the state space are finite. Given the fact that also the time considered for a life insurance is finite, the required property follows.

Theorem 11.1.14 *For* $x \in \mathcal{X}$, *as defined above we have the following:*

1. $E[\Delta N_{ij}(s)|X_t = k] = p_{ki}(t, s)p_{ij}(s, s+1)$,
2. $E[I_i(s)|X_t = k] = p_{ki}(t, s)$,
3. $E[x_s|X_t = k] =$

$$\sum_{(i,j)\in S^2} p_{ki}(t, s-1)p_{ij}(s-1, s)a_{ij}^{Post}(s-1) + \sum_{i\in S} p_{ki}(t, s)a_i^{Pre}(s),$$

and we assume that $p_{ki}(t, s-1) = 0$ *if* $t = s$.

Proof We leave the proof of this proposition to the reader as an exercise.

Definition 11.1.15 The abstract vector space of financial instruments we denote by $\mathcal{Y}$. Elements of this vector space are for example all zero coupon bonds, shares, options on shares etc.

Remark 11.1.16 • Link to the arbitrage free pricing theory: If we assume that Q does not allow arbitrage we are in the set up of Chap. 8. In Proposition 8.2.15 we have seen that $\pi(X) = E^Q[\beta_T X]$, where β_T denotes the risk free discount rate. In the context of the above, we would have $\pi_0(x) = Q[x] = E^P[\phi_T x]$. Hence we can identify $\phi_T = \frac{dQ}{dP}\beta_T$. In consequence we can interpret a deflator as a discounted Radon-Nikodym density with respect to the two measures P and Q.
• In the same sense the concepts of Definition 11.1.11 have a lot in common with the definition of the present values of a cash flow stream as defined in Chap. 4.6.
• For the interest rate model in Sect. 9.4 we know that

$$B_t(s) = \pi_t\left(\mathcal{Z}_{(s)}\right)$$
$$= E\left[\exp(-\int_t^s r_u du) \times \exp(-\int_t^s \lambda^T dW - \frac{1}{2}\int_t^s \lambda^T \lambda\, du) \mid \mathcal{G}_t\right],$$

and hence we have actually calculated the corresponding deflators as follows:

$$\phi_t(s) = \exp(-\int_t^s r_u du) \times \exp(-\int_t^s \lambda^T dW - \frac{1}{2}\int_t^s \lambda^T \lambda\, du).$$

Theorem 11.1.17 *Let Q be a positive, continuous functional $Q : \mathcal{X} \to \mathbb{R}$, and assume $Q[\bullet] = <\phi, \bullet>$, with $\phi = (\phi_t)_{t\in\mathbb{N}}$ $\mathbb{F}$-adapted. In this case $(\phi_t Q_t[x])_{t\in\mathbb{N}}$ is an $\mathbb{F}$-martingale over P.*

Proof Since $\mathcal{F}_t \subset \mathcal{F}_{t+1}$ and the projection property of the conditional expectation we have

$$E^P[\phi_{t+1} Q_{t+1}[x]|\mathcal{F}_t] = E^P[E^P[\sum_{k\in\mathbb{N}} \phi_k x_k\, Q_{t+1}[x]|\mathcal{F}_{t+1}]|\mathcal{F}_t]$$
$$= E^P[\sum_{k\in\mathbb{N}} \phi_k x_k\, Q_{t+1}[x]|\mathcal{F}_t]$$
$$= \phi_t Q_t[x].$$

Example 11.1.18 (Replicating Portfolio Mortality) In this first example we consider a term insurance, for a 50 year old man with a term of 10 years, and we assume that this policy is financed with a regular premium payment. Hence there are actually two different payment streams, namely the premium payment stream and the benefits payment stream. For sake of simplicity we assume that the yearly mortality is $(1 + \frac{x-50}{10} \times 0.1)\%$. We assume that the death benefit amounts to 100,000 USD and we assume that the premium has been determined with an interest rate $i = 2\%$. In this case the premium amounts to $P = 1394.29$. The replicating portfolio in the sense of expected cash flows at inception is therefore given as follows (cf Proposition 11.1.14). We remark that the units have been valued with two (flat) yield curves with interest rates of 2 and 4% respectively, and remark the use of arbitrary yield curves does not imply additional complexity.

Age	Unit	Units for mortality	Units for premium	Total units	Value $i = 2\%$	Value $i = 4\%$
50	$\mathcal{Z}_{(0)}$	–	−1394.28	−1394.28	−1394.28	−1394.28
51	$\mathcal{Z}_{(1)}$	1000.00	−1380.34	−380.34	−372.88	−365.71
52	$\mathcal{Z}_{(2)}$	1089.00	−1365.16	−276.16	−265.43	−255.32
53	$\mathcal{Z}_{(3)}$	1174.93	−1348.77	−173.84	−163.81	−154.54
54	$\mathcal{Z}_{(4)}$	1257.56	−1331.24	−73.67	−68.06	−62.97
55	$\mathcal{Z}_{(5)}$	1336.69	−1312.60	24.09	21.82	19.80
56	$\mathcal{Z}_{(6)}$	1412.12	−1292.91	119.20	105.85	94.21
57	$\mathcal{Z}_{(7)}$	1483.67	−1272.23	211.44	184.07	160.67
58	$\mathcal{Z}_{(8)}$	1551.18	−1250.60	300.57	256.54	219.62
59	$\mathcal{Z}_{(9)}$	1614.50	−1228.09	386.41	323.33	271.48
60	$\mathcal{Z}_{(10)}$	1673.52	–	1673.52	1372.87	1130.57
Total					**0.00**	**−336.47**

Exercise 11.1.19 (Replicating Portfolio Disability) Consider a disability cover and calculate the replicating portfolios for a deferred disability annuity and a disability in payment.

11.2 Cost of Capital

In Sect. 11.1 we have seen how to abstractly valuate $x \in \mathcal{X}$ by means of a pricing functional Q. For some financial instruments $y \in \mathcal{Y}^\star$ we can directly observe $Q[y]$ such as for a lot of zero coupons bonds $\mathcal{Z}_{(\bullet)}$. On the other hand this is not always possible.

Definition 11.2.1 We denote by $\mathcal{Y}^\star$ the set of all stochastic cash flows in $x \in \mathcal{Y}$ such that $Q[x]$ is observable. With $\tilde{\mathcal{Y}} = \mathrm{span} < \mathcal{Y}^\star >$ we denote the vector space generated by $\mathcal{Y}^\star$ and we define:

1. $x \in \mathcal{Y}^\star$ is called of level 1.
2. $x \in \tilde{\mathcal{Y}}$ is called of level 2.
3. $x \in \mathcal{Y} \setminus \tilde{\mathcal{Y}}$ is called of level 3.

Remark 11.2.2 It is clear that the model uncertainty and the difficulties to value assets or liabilities increases from level 1 to level 3. Since we are interested in market values only the valuation of level 1 assets and liabilities are really reliable. For level 2 assets and liabilities on has to find a sequence of $x_n = \sum_{k=1}^{n} \alpha_k e_k$ with $e_k \in \mathcal{Y}^\star$ such that $x = \lim_{n\to\infty} x_n$. Since we *assume* that Q is linear and continuous we can calculate

$$
\begin{aligned}
Q[x] &= \lim_{n\to\infty} Q[x_n] \\
&= \lim_{n\to\infty} \sum_{k=1}^{n} Q[\alpha_k e_k] \\
&= \lim_{n\to\infty} \sum_{k=1}^{n} \alpha_k Q[e_k].
\end{aligned}
$$

For level 3 assets and liabilities the situation is even more difficult, since there is no obvious way to do it. The best, which we can be done is to *define* $\tilde{Q}[x]$ such that $\tilde{Q}[x] = Q[x] \forall x \in \mathcal{Y}$ and *hope* that $\tilde{Q}[x] \approx Q[x]$ for the $x \in \mathcal{Y}$ we want to valuate. In most cases such $\tilde{Q}[\bullet]$ are based on first economic principles. In the following we want to see how the *Cost of Capital* concept works for insurance liabilities and how we can concretely implement it.

Definition 11.2.3 (*Utility Assumption*) If we have $x, y \in L^2(\Omega, \mathcal{A}, P)^+$, with $x = E[y]$. A rational investor would normally prefer x, since there is less uncertainty. The way to understand this, is by using utility functions. For $x \in L^2(\Omega, \mathcal{A}, P)^+$ and u a concave function, the utility of x is defined as $E[u(x)]$. The idea behind utilities is that the first 10,000 USD are higher valued than the one 10,000 USD from 100,000 USD to 110,000 USD. Hence the increase of utility per fixed amount decreases if amounts increase. As a consequence of the Jensen's inequality, we see that the utility of a constant amount is higher than the utility of a random payout with the same expected value.

Definition 11.2.4 Let $x = (x_k)_{k \in \mathbb{N}} \in \mathcal{X}$ be an insurance cash flow, for example generated by a Markov model.

1. In this case we define the expected cash flows by

$$CF(x) = (E[x_k])_{k \in \mathbb{N}}.$$

2. The corresponding portfolio of financial instruments in the vector space $\mathcal{Y}$ we define by

$$VaPo^{CF}(x) = \sum_{k \in \mathbb{N}} CF(x)_k \mathcal{Z}_{(k)} \in \mathcal{Y}$$

3. By $R(x)$ we denote the residual risk portfolio given by

$$R(x) = x - VaPo^{CF}(x)$$
$$= \sum_{k \in \mathbb{N}} (x_k - CF(x)_k) \mathcal{Z}_{(k)} \in \mathcal{Y}$$

4. For a given $x \in \mathcal{X}$ we denote by $VaPo^*(x)$ an approximation $y \in \tilde{\mathcal{Y}}$ of x, such that $||x - VaPo^*(x)|| \leq ||x - VaPo^{CF}(x)||$.

Since we are sometimes interested in conditional expectations, we will also use the following notations for $A \in \mathcal{A}$:

$$CF(x \,|\, A) = (E[x_k \,|\, A])_{k \in \mathbb{N}},$$
$$VaPo^{CF}(x \,|\, A) = \sum_{k \in \mathbb{N}} CF(x \,|\, A)_k \mathcal{Z}_{(k)} \in \mathcal{Y},$$

Theorem 11.2.5 *The value of $x \in \mathcal{X}$ can be decomposed in*

$$Q[x] = Q[VaPo^{CF}(x)] + Q[R(x)],$$

and we have

$$Q[VaPo^{CF}(x)] \geq Q[x]$$

if we use the utility assumption.

Remark 11.2.6   1. We will denote $x \in \mathcal{X}$ with $x \leq 0$ as a liability. Proposition 11.2.5 hence tells us that we need to reserve more than $Q[VaPo^{CF}(x)]$ for this liability as a consequence of the corresponding uncertainty.

   2. A risk measure is a functional (not necessarily linear) $\psi : \mathcal{X} \to \mathbb{R}$ which aims to measure the capital needs in an adverse scenario. There are two risk measures, which are commonly used the *Value at Risk* and the *Expected Shortfall* to a given quantile $\alpha \in \mathbb{R}$. The value at risk (VaR) is defined as the corresponding quantile minus the expected value. The expected shortfall is the conditional expectation of the random variable given a loss bigger than the corresponding loss, again minus the expected value. We can hence speak about a 99.5% VaR or a 99% expected shortfall. It is worthwhile to remark that these two concepts are normally applied to losses. Hence in the context introduced above one would strictly speaking calculating the VaR$(-x)$, when considering $x \in \mathcal{X}$. Furthermore in a lot of applications, such as Solvency II, we assume that there is a Dirac measure (aka stress scenario), which just represents the corresponding VaR-level for example. So concretely the stress scenarios, which are used under Solvency II should in principle represent the corresponding point (Dirac) measures at to the confidence level 99.5 %. In the concrete set up, one would for example assume that $q_x(\omega) \in L^2(\Omega, \mathcal{A}, P)$ is a stochastic mortality and one would define the $A, B \in \mathcal{A}$, as the corresponding probabilities in the average and in the tail. In consequence for a policy $x \in \mathcal{X}$, we would have two replicating portfolios, namely $VaPo^{CF}(x \mid A)$ for the average and $VaPo^{CF}(x \mid B)$ for the stressed event according to the risk measure chosen. The corresponding required risk capital is then given (in present value terms) by $Q[VaPo^{CF}(x \mid B) - VaPo^{CF}(x \mid A)]$.

Definition 11.2.7 (*Required Risk Capital*) For a risk measure ψ_α such as VaR or expected shortfall to a security level α we define the required risk capital at time $t \in \mathbb{N}$ by

$$RC_t(x) = \psi_\alpha(p_k(x - VaPo^{CF}(x))).$$

Remark 11.2.8   1. If we use $VaR_{99.5\%}$ the required risk capital at time t corresponds to the capital needed to withstand a 1 in 200 year event.

   2. The definition above could apply to individual insurance policies, but is normally applied to insurance portfolios $\tilde{x} = \sum_{k=1}^{n} x_k$, where x_k are the individual

insurance policies. As we have seen in Sect. 9.3 the pure diversifiable risk disappears for $n \to \infty$.

3. What is more material than the diversifiable risk is the risk, which affects all of the individual insurance policies at the same time, such as a pandemic event, where the overall mortality could increase by 1 % in a certain year such as 1918.

Definition 11.2.9 (*Cost of Capital*) For a unit cost of capital $\beta \in \mathbb{R}^+$ and an insurance portfolio $\tilde{x} \in \mathcal{X}$, we define:

1. The present value of the required risk capital by

$$PVC(\tilde{x}) = Q[\sum_{k \in \mathbb{N}} RC_t(\tilde{x}) \mathcal{Z}_{(k)}].$$

2. The cost of capital $CoC(\tilde{x})$ is given by:

$$CoC(\tilde{x}) = \beta \times PVC(\tilde{x}),$$

and $\tilde{Q}$ is defined by $\tilde{Q}[\tilde{x}] = Q[VaPo^{CF}(\tilde{x})] + \beta PVC(\tilde{x})$.

Remark 11.2.10 1. The concept as defined before is somewhat simplified, since one normally assumes that the required capital C from the shareholder is $\alpha \times C$ after tax and investment income on capital, Assume a tax-rate κ and a risk-free yield of i. In this case we have

$$\alpha \times C = i \times (1 - \kappa) \times C + \beta \times C,$$

and hence $\beta = \alpha - i \times (1 - \kappa)$. In reality the calculation can still become more complex since we discount future capital requirements risk-free and because of the fact that the interest rate i is not constant. In order to avoid these technicalities, we will assume for this book that i is constant.

2. We remark $\tilde{Q}[\tilde{x}]$ is not uniquely determined, but depends on a lot of assumptions such as ψ_α, α, β, ...

3. For the moment we did not yet see how to actually model $\tilde{x}$ and we remark that one is normally focusing on the non-diversifiable part of the risks within $\tilde{x}$.

Example 11.2.11 We continue with Example 11.1.18 and we assume that the risk capital is given by a pandemic event where $\Delta q_x = 1\%$ for all ages. This roughly corresponds to the increase in mortality of 1918 as a consequence of the Spanish flu pandemic. The aim of this example is to calculate the required risk capital and the market value of this policy based on the cost of capital method using $\beta = 6\%$. The required risk capital in this context can be calculated as $\Delta q_x \times 100,000$ and we get the following results:

Age	Unit	Units for risk capital	Units for benefits	Total units	$-\tilde{Q}[x]$ $i = 2\%$	$-\tilde{Q}[x]$ $i = 4\%$
50	$Z_{(0)}$	1000.00	−1394.28	−1334.28	−1334.28	−1334.28
51	$Z_{(1)}$	990.00	−380.34	−320.94	−314.65	−308.60
52	$Z_{(2)}$	979.11	−276.16	−217.41	−208.97	−201.01
53	$Z_{(3)}$	967.36	−173.84	−115.80	−109.12	−102.95
54	$Z_{(4)}$	954.78	−73.67	−16.38	−15.14	−14.00
55	$Z_{(5)}$	941.41	24.09	80.57	72.98	66.22
56	$Z_{(6)}$	927.29	119.20	174.84	155.25	138.18
57	$Z_{(7)}$	912.45	211.44	266.19	231.73	202.28
58	$Z_{(8)}$	896.94	300.57	354.39	302.47	258.95
59	$Z_{(9)}$	880.80	386.41	439.26	367.55	308.61
60	$Z_{(10)}$	–	1673.52	1673.52	1372.87	1130.57
Total					**520.69**	**143.98**

We remark that the value of the policy at inception becomes positive, which means nothing else, that the insurance company does need equity capital to cover the economic loss. It is obvious that this is the case for $i = 2\%$, since the premium principle did not allow for a compensation of the risk capital. More interestingly even at the higher interest rate the compensating effect is not big enough to turn this policy into profitability.

Exercise 11.2.12 In the same sense as for the mortality example calculate the respective risk capitals and the $\tilde{Q}$ for a disability cover.

11.3 Inclusion in the Markov Model

In this section we want to have a look how we could concretely use the recursion technique for the calculation of the cost of capital in a Markov chain similar environment. In order to do that we look at an insurance policy with a term of 1 year.

We assume that we have a mortality of q_x in case of a "normal" year with a probability of $(1 - \alpha)$ and an excess mortality of Δq_x in an extreme year with probability α. We denote with $\Gamma = \frac{q_x + \Delta q_x}{q_x}$. Furthermore we assume a mortality benefit of 100,000. In this case we get the following by some simple calculations:

$$VaPo^{CF}(x) = (\delta_{1k}(q_x + \alpha(\Gamma - 1)q_x \times 100000))_{k \in \mathbb{N}}$$
$$RC_1(x) = (\delta_{1k}(1 - \alpha)(\Gamma - 1)q_x \times 100000)_{k \in \mathbb{N}}$$
$$\tilde{Q}[x] = Q[(\delta_{1k}(q_x + \alpha(\Gamma - 1)q_x \times 100000 + \\ + \beta(1 - \alpha)(\Gamma - 1) \times 100000))_{k \in \mathbb{N}}]$$

We see that the price of this insurance policy with only payments at time 1 can be decomposed into a part representing best estimate mortality:

$$\delta_{1k}\{q_x\,(1 + \alpha\,(\Gamma - 1))\},$$

where we can arguably say that this $\tilde{q}_x = q_x\,(1 + \alpha\,(\Gamma - 1))$ is our actual best-estimate mortality. On top of that we get a charge for the excess mortality Δq_x with an additional cost of β. Hence we get the following:

1. There is a contribution to the reserve from the people surviving the year with a probability p_x.
2. There is a contribution to the reserve from the people dying in normal years with probability q_x and the defined benefit $a_{*\dagger}^{\text{post}}$, and
3. There is finally a contribution of the people dying in extreme years with probability Δq_x and the additional cost of defined benefit of $\beta \times a_{*\dagger}^{\text{post}}$.

The interesting fact is that we can actually use the same recursion of the reserves for the Markov chain model as in Proposition 4.7.3 with the exception that now the "transition probabilities" do not fulfil anymore the requirement that their sum equals 1. However this method provides a pragmatic way to implement the cost of capital in legacy admin systems.

The main problem for the determining of the corresponding Markov chain model is the underlying stochastic mortality model. For the QIS 5 longevity model a similar calculation can be used. In this model it is assumed that the mortality drops by 25 % in an extreme scenario. Hence the calculation goes along the following process:

1. Determine $x_1 = VaPo^{CF}(\tilde{x})$.
2. Determine $x_2 = VaPo^{CF}(\tilde{x})$ for stressed mortality.
3. $\tilde{Q}[x] = Q[x_1] + \beta\,Q[x_2 - x_1]$

Example 11.3.1 In this example we want to revisit the Exercise 11.1.18 and we want again to calculate the market value of the insurance liability, but this time with the recursion. We get the following results:

Age	Benefit normal	Benefit premium	Excess risk	Math Res. $i = 2\%$	Value $i = 2\%$	Value $i = 4\%$
50	100000	−1394.28	6000	**0.00**	**520.69**	**143.98**
51	100000	−1394.28	6000	426.43	901.09	542.82
52	100000	−1394.28	6000	765.56	1193.21	861.67
53	100000	−1394.28	6000	1015.22	1394.79	1096.96
54	100000	−1394.28	6000	1172.95	1503.20	1244.68
55	100000	−1394.28	6000	1235.88	1515.45	1300.33
56	100000	−1394.28	6000	1200.79	1428.16	1258.89
57	100000	−1394.28	6000	1064.00	1237.49	1114.74
58	100000	−1394.28	6000	821.42	939.18	861.64
59	100000	−1394.28	6000	468.45	528.45	492.63
60				0	0	0

We remark that this calculation was much faster to calculate since it is based on Thiele's difference equation for the mathematical reserves, and we get at the same time the corresponding results for the classical case and also for the case using the cost of capital approach.

As seen in the calculation above there is a small second order effect, which we can detect, when looking more closely. The results below correspond to the 2% valuation:

Direct Method   520.698380872792
Recursion   520.698380872793

Exercise 11.3.2 Perform the corresponding calculation for the disability example.

11.4 Asset Liability Management

Until now we have looked only at insurance liabilities as an $x \in \mathcal{X}$. An insurance company needs to cover its insurance liabilities $l = \sum x_\iota \in \mathcal{X}$ with corresponding assets, which are also elements in $\mathcal{X}$.

Definition 11.4.1 (*Assets and Liabilities*) An $x \in \mathcal{X}$ with a valuation functional Q is called

1. an asset if $Q[x] \geq 0$ and
2. a liability if $Q[x] \leq 0$.

Definition 11.4.2 (*Insurance balance sheet*) An insurance balance sheet consists of a set of assets $(a_i)_{i \in I}$ and a set of liabilities $(l_j)_{j \in J}$. The equity of an insurance balance sheet is defined as

$$e = \sum_{i \in I} a_i + \sum_{j \in J} l_j.$$

The insurance entity is called bankrupt if $Q[e] < 0$.

Definition 11.4.3 In an insurance market, each insurance company is required to hold an adequate amount of risk capital in order to absorb shocks. In order to do that, the regulator defines a risk measure ψ_α to a security level α. In this context an insurance company is called solvent if:

$$Q[e] \geq \psi_\alpha(e).$$

Remark 11.4.4 Note that an insurance regulator may not want to use a market consistent approach. Never the less the above definition can be used, be suitably adjust ψ.

Definition 11.4.5 (*Asset Liability Management*) Under Asset Liability Management we understand the process of analysing $(l_j)_{j \in J}$ and the (dynamic) management of $(a_i)_{i \in I}$ in order to achieve certain targets , such as remaining solvent.

Definition 11.4.6 For an insurance liability $l \in \mathcal{X}$ an asset portfolio $(a_i)_{i \in I}$ is called:

1. *matching* if $\sum_{i \in I} a_i + l = 0$, and
2. *cash flow matching* if $\sum_{i \in I} a_i + VaPo^{CF}(l) = 0$.

Remark 11.4.7 We remark that is normally not feasible to do a perfect matching, and hence one normally uses a cash flow matching to a achieve a proxy for a perfect match. We also remark that in this case the shareholder equity needs still be able to absorb the basis risk $l - VaPo^{CF}(l)$.

Definition 11.4.8 (*Duration*) The duration for an $x \in \mathcal{X}$ with $x = \sum_{i \in \mathbb{N}} \alpha_i \, \mathcal{Z}_{(i)}$ and $\alpha_i \geq 0$ is defined by

$$d(x) = \frac{Q[\sum_{i \in \mathbb{N}} \alpha_i \times i \times \mathcal{Z}_{(i)}]}{Q[\sum_{i \in \mathbb{N}} \alpha_i \times \mathcal{Z}_{(i)}]}$$

We say that an asset portfolio $(a_i)_{i \in I}$ is duration matching a liability l if the following two conditions are fulfilled:

1. $Q[\sum_{i \in I} a_i + l] = 0$, and
2. $d(\sum_{i \in I} a_i) = d(-l)$.

Example 11.4.9 In this example we want to further elaborate on the Example 11.1.18 and we want to see how the replicating scenario changes in case a pandemic occurs in year three, with an excess mortality of 1 %. We want also to have a look on what risk is implied in this, assuming that the pandemic at the same time leads to a reduction of interest rates down from 2 to 0.5 %. Finally we want to see an example how we could do a perfect cash flow matching portfolio and duration matched portfolio.

Definitions We assume that $A \in \mathcal{A}$ represents the information that we have going to have average mortality after year 3 and three and that the person survived until then (year 2). In the same sense we assume that $B \in \mathcal{A}$ represents the same as A but with the exception that we assume a pandemic event in the year 3 with an average excess mortality of 1%. For simplicity reasons (to avoid notation) we use $x, y \in \mathcal{X}$ as abbreviations for the corresponding conditional random variables.

Calculation of the Replicating Portfolios In a first step we will calculate the replicating portfolios (starting at time 2) with respect to both A and B. Doing this we get the following results for case A:

For case B we get:

We note two things:

- The pandemic happens when the person is aged 53 and we see the impact in $\mathcal{Z}_{(2)}$ at age 54. This has to do with the convention that we assume that the deaths occur at the end on the year, hence just before the person gets 54.
- We see that the difference in reserves amounts to $1704.05 - 765.56 = 938.49$ which represents the economic loss as a consequence of the pandemic. The biggest contributor to this loss is the increased death benefit, e.g. $926.52 - 1284.40 = 962.87$.

Age	Unit	Units for mortality	Units for premium	Total units	Value $i = 2\%$	Value $i = 4\%$
52	$\mathcal{Z}_{(0)}$	–	−1394.28	−1394.28	−1394.28	−1394.28
53	$\mathcal{Z}_{(1)}$	1200.00	−1377.55	−177.55	−174.07	−170.72
54	$\mathcal{Z}_{(2)}$	1284.40	−1359.64	−75.24	−72.32	−69.57
55	$\mathcal{Z}_{(3)}$	1365.21	−1340.61	24.60	23.18	21.87
56	$\mathcal{Z}_{(4)}$	1442.25	−1320.50	121.75	112.48	104.07
57	$\mathcal{Z}_{(5)}$	1515.33	−1299.37	215.95	195.59	177.49
58	$\mathcal{Z}_{(6)}$	1584.27	−1277.28	306.99	272.59	242.61
59	$\mathcal{Z}_{(7)}$	1648.95	−1254.29	394.65	343.57	299.90
60	$\mathcal{Z}_{(8)}$	1709.23	–	1709.23	1458.81	1248.91
Total					**765.56**	**460.30**

Age	Unit	Units for mortality	Units for premium	Total units	Value $i = 2\%$	Value $i = 4\%$
52	$\mathcal{Z}_{(0)}$	–	−1394.28	−1394.28	−1394.28	−1394.28
53	$\mathcal{Z}_{(1)}$	1200.00	−1377.55	−177.55	−174.07	−170.72
54	$\mathcal{Z}_{(2)}$	**2272.40**	−1345.87	**926.52**	890.54	856.62
55	$\mathcal{Z}_{(3)}$	1351.38	−1327.03	24.35	22.95	21.65
56	$\mathcal{Z}_{(4)}$	1427.64	−1307.12	120.51	111.34	103.01
57	$\mathcal{Z}_{(5)}$	1499.97	−1286.21	213.76	193.61	175.70
58	$\mathcal{Z}_{(6)}$	1568.22	−1264.34	303.88	269.83	240.16
59	$\mathcal{Z}_{(7)}$	1632.24	−1241.58	390.65	340.09	296.86
60	$\mathcal{Z}_{(8)}$	1691.91	–	1691.91	1444.03	1236.26
Total					**1704.05**	**1365.28**

Matching asset portfolios Based on the above it is now easy to calculate the cash flow matching portfolio, by just investing the different amounts of liabilities into the corresponding assets, such as buying $24.60\mathcal{Z}_{(3)}$. We remark that consequently we would have to sell $-177.55\mathcal{Z}_{(1)}$. In normal circumstances for mature businesses this will not occur, since it is a consequence that we consider a term insurance policy and not for example an endowment.

Mismatch in case of a pandemic The table below finally shows the cash flow mismatch as a consequence of the pandemic and we see that in this case the present values do not have a big impact since the main difference is at time 1.

Age	Unit	Units normal	Units stress	Difference units	Value $i = 2\%$	Value $i = 0\%$
52	$\mathcal{Z}_{(0)}$	−1394.28	−1394.28	0.00	0.00	0.00
53	$\mathcal{Z}_{(1)}$	−177.55	−177.55	0.00	0.00	0.00
54	$\mathcal{Z}_{(2)}$	−75.24	926.52	1001.77	962.87	1001.77
55	$\mathcal{Z}_{(3)}$	24.60	24.35	−0.24	−0.23	−0.24
56	$\mathcal{Z}_{(4)}$	121.75	120.51	−1.23	−1.13	−1.23
57	$\mathcal{Z}_{(5)}$	215.95	213.76	−2.18	−1.98	−2.18
58	$\mathcal{Z}_{(6)}$	306.99	303.88	−3.11	−2.76	−3.11
59	$\mathcal{Z}_{(7)}$	394.65	390.65	−3.99	−3.48	−3.99
60	$\mathcal{Z}_{(8)}$	1709.23	1691.91	−17.31	−14.78	−17.31
Total					**938.49**	**973.67**

Example 11.4.10 (Lapses) In this example we want to see how lapses can influence the replicating portfolios. In order to do that we have to change the Example 11.1.18 a little bit, as follows:

- We consider a term insurance, for a 50 year old man with a term of 10 years, and we assume that this policy is financed with a regular premium payment. Hence there are actually two different payment streams, namely the premium payment stream and the benefits payment stream. For sake of simplicity we assume that the yearly mortality is $(1 + \frac{x-50}{10} \times 0.1)\%$. We assume that the benefit amounts to 100.000 USD and we assume that the premium has been determined with an interest rate $i = 2\%$.
- In this case the premium amounts to $P = 9562.20$.
- In addition the policyholder can surrender the policy at any time and gets back 98 % of the expected future cash flows valued at the pricing interest rate of 2%. We remark here that this is a risk since the surrenders can happen in case the market value of the corresponding units is below the surrender value.
- We remark that the units have been valued with two (flat) yield curves with interest rates of 2 and 4% respectively.

In order to calculate this example we will perform the following steps:

1. Calculation of the cash flow matching portfolio in case of no surrenders.
2. Calculation of the cash flow including lapses with an average lapse rate of 7 %
3. Calculation of the cash flows at time 2, assuming average lapses, lapses at 25 % at time 2.

Calculation of the cash flow matching portfolio in case of no surrenders

Age	Unit	Units for mortality	Units for premium	Total units	Value $i = 2\%$	Value $i = 4\%$
50	$Z_{(0)}$	–	−9562.20	−9562.20	−9562.20	−9562.20
51	$Z_{(1)}$	1000.00	−9466.57	−8466.57	−8300.56	−8140.94
52	$Z_{(2)}$	1089.00	−9362.44	−8273.44	−7952.17	−7649.26
53	$Z_{(3)}$	1174.93	−9250.09	−8075.16	−7609.40	−7178.79
54	$Z_{(4)}$	1257.56	−9129.84	−7872.27	−7272.76	−6729.25
55	$Z_{(5)}$	1336.69	−9002.02	−7665.32	−6942.72	−6300.34
56	$Z_{(6)}$	1412.12	−8866.99	−7454.87	−6619.71	−5891.69
57	$Z_{(7)}$	1483.67	−8725.12	−7241.45	−6304.11	−5502.90
58	$Z_{(8)}$	1551.18	−8576.79	−7025.61	−5996.29	−5133.54
59	$Z_{(9)}$	1614.50	−8422.41	−6807.90	−5696.55	−4783.14
60	$Z_{(10)}$	–	88080.30	88080.30	72256.53	59503.90
Total					**0**	**−7368.19**

We remark that the there is considerable value in the policy if we assume no lapses, in case we earn a higher interest rate, such as 4 %.

Calculation of the cash flow matching portfolio in case of 7% surrenders

Age	Unit	Units for mortality	Units for premium	Total units	Value $i = 2\%$	Value $i = 4\%$
50	$\mathcal{Z}_{(0)}$	–	−9562.20	−9562.20	−9562.20	−9562.20
51	$\mathcal{Z}_{(1)}$	1019.67	−8797.22	−7777.54	−7625.04	−7478.40
52	$\mathcal{Z}_{(2)}$	1594.01	−8084.65	−6490.63	−6238.59	−6000.95
53	$\mathcal{Z}_{(3)}$	2066.98	−7421.70	−5354.72	−5045.87	−4760.33
54	$\mathcal{Z}_{(4)}$	2449.23	−6805.70	−4356.47	−4024.71	−3723.93
55	$\mathcal{Z}_{(5)}$	2750.73	−6234.02	−3483.29	−3154.92	−2863.01
56	$\mathcal{Z}_{(6)}$	2980.77	−5704.13	−2723.35	−2418.26	−2152.31
57	$\mathcal{Z}_{(7)}$	3147.94	−5213.57	−2065.63	−1798.25	−1569.71
58	$\mathcal{Z}_{(8)}$	3260.18	−4759.99	−1499.81	−1280.07	−1095.89
59	$\mathcal{Z}_{(9)}$	3324.77	−4341.11	−1016.34	−850.42	−714.06
60	$\mathcal{Z}_{(10)}$	5486.26	42159.27	47645.53	39085.93	32187.61
Total					**−2912.44**	**−7733.21**

We remark that at that time, the company makes still some additional gains as a consequence of the 2% surrender penalty. **Calculation of the cash flow matching portfolio in case of high surrenders:** We assume that there has been observed an exceptional lapse rate at time 2 of 25% of the portfolio.

Age	Unit	Units for mortality	Units for premium	Total units	Value $i = 2\%$	Value $i = 4\%$
50	$\mathcal{Z}_{(0)}$	–	−9562.20	−9562.20	−9562.20	−9562.20
51	$\mathcal{Z}_{(1)}$	1019.67	−8797.22	−7777.54	−7625.04	−7478.40
52	$\mathcal{Z}_{(2)}$	1594.01	−8084.65	−6490.63	−6238.59	−6000.95
53	$\mathcal{Z}_{(3)}$	4773.16	−5966.47	−1193.30	−1124.48	−1060.84
54	$\mathcal{Z}_{(4)}$	1968.98	−5471.25	−3502.26	−3235.55	−2993.75
55	$\mathcal{Z}_{(5)}$	2211.37	−5011.66	−2800.29	−2536.31	−2301.63
56	$\mathcal{Z}_{(6)}$	2396.30	−4585.67	−2189.36	−1944.09	−1730.28
57	$\mathcal{Z}_{(7)}$	2530.70	−4191.30	−1660.60	−1445.65	−1261.92
58	$\mathcal{Z}_{(8)}$	2620.93	−3826.66	−1205.73	−1029.08	−881.01
59	$\mathcal{Z}_{(9)}$	2672.85	−3489.91	−817.05	−683.67	−574.05
60	$\mathcal{Z}_{(10)}$	4410.52	33892.74	38303.27	31422.02	25876.31
Total					**−4002.67**	**−7968.76**

ALM Risk of mass lapses Finally we want to look what happens when we have mass lapses as indicated before, but if we have invested in the cash flow matching portfolio according to average 7 % lapses. Hence we have to calculate the assets according to 7 % lapses and the liabilities according 25 % lapses.

Now we see that the lapses induce quite a big risk for the company since they lose in case of mass lapses almost 1 % of the face value of the policy, more concretely $1134.26 - 254.76 = 879.50$.

The above example shows very clearly how the behaviour of the policyholders can change the cash flow matching portfolio and in consequence induces a risk. As a consequence the risk minimising portfolio in the sense of $VaPo^*(x)$ for an insurance

Age	Unit	Units for assets	Units for liability	Total units	Value $i = 2\%$	Value $i = 4\%$
52	$\mathcal{Z}_{(0)}$	−6490.63	6490.63	0	0	0
53	$\mathcal{Z}_{(1)}$	−5354.72	1193.30	−4161.42	−4079.82	−4001.36
54	$\mathcal{Z}_{(2)}$	−4356.47	3502.26	−854.21	−821.04	−789.76
55	$\mathcal{Z}_{(3)}$	−3483.29	2800.29	−682.99	−643.60	−607.18
56	$\mathcal{Z}_{(4)}$	−2723.35	2189.36	−533.99	−493.32	−456.45
57	$\mathcal{Z}_{(5)}$	−2065.63	1660.60	−405.02	−366.84	−332.90
58	$\mathcal{Z}_{(6)}$	−1499.81	1205.73	−294.08	−261.13	−232.41
59	$\mathcal{Z}_{(7)}$	−1016.34	817.05	−199.28	−173.48	−151.43
60	$\mathcal{Z}_{(8)}$	47645.53	−38303.27	9342.26	7973.53	6826.29
Total					**1134.26**	**254.76**

portfolio $x \in \mathcal{X}$ does also consist of additional assets offsetting the corresponding risks. In the above example the corresponding asset would be a (complex) put option, which allows to sell the bond portfolio at the predefined (book-) values. So in reality insurance companies aim to model these risk in order to determine the corresponding assets to manage and reduce the undesired risk.

In the example above we have assumed that at a given year 25% of the policies in force lapse. In practise one models the dynamic lapse behaviours. Eg the lapse rate is a function of the interest differential between market and book yields. Normally the corresponding lapse rates stay below 1, which is interesting. Assuming a market efficient behaviour, one would expect that there is a binary decision of the policyholders to stick to the contract or to lapse as a function of the before mentioned interest differential. In consequence the underlying theory how to model such policyholder behaviour is not as crisp and transparent as with the arbitrage free pricing theory, since market efficient behaviours is normally not observed. As a corollary there is a lot of model risk intrinsic to these calculations and it is important to test the results from the models with different scenarios.

Remark 11.4.11 At the end of this section a remark on how to determine a $VaPo^*(x)$ for an $x \in \mathcal{X}$: One normally models an $l \in \mathcal{X}$ and simulates $l(\omega)$ together with some test assets $D \subset \mathcal{Y}$ observable prices and cash flows. We denote $D = \{d_1, \ldots d_n\}$. Hence at the end of this process we have a vector

$$\mathcal{W} := (l(\omega_i), d_1(\omega_i), \ldots d_n(\omega_i))_{i \in I}.$$

Now the process is quite canonical:

1. We define a distance between two $x, y \in \mathcal{X}$, for example by means of $||x||$ as defined.
2. We solve the numerical optimisation problem, for minimising the distance between l and the target $y \in$ span $< \mathcal{D} >$.

We note two things:

- The numerical procedures to determine y can sometimes prove to be difficult since the corresponding design matrix can be near to a singular matrix, and hence additional care is needed.
- In case of the $\| \bullet \|$ defined before, we remark that it has been deducted from the Hilbert space $\mathcal{X}$. Hence what we actually doing is to use the projection $\tilde{p} : \mathcal{X} \to$ span $< \mathcal{D} >$, which can be expressed by means of $< \bullet, \bullet >$. We remark that $y = \tilde{p}(x)$.

Chapter 12
Policyholder Bonus Mechanism

The aim of this chapter is to introduce the concept of policyholder bonus and the corresponding effects.

12.1 Concept of Surplus: Traditional Approach and Legal Quote

Lets reconsider example 11.4.10. We have used in this context a interest rate for pricing of 2%. According to the EU laws the technical interest rate must normally not exceed 60% of the average government bond yields. This would mean that at that time the average government bond yield was above 3.3%—assume 4%. If we redo the corresponding calculations, we get the following:

Age	Unit	Units for mortality	Units for premium	Total units	Value $i = 2\%$	Value $i = 4\%$
50	$Z_{(0)}$	–	−9562.20	−9562.20	−9562.20	−9562.20
51	$Z_{(1)}$	1000.00	−9466.57	−8466.57	−8300.56	−8140.94
52	$Z_{(2)}$	1089.00	−9362.44	−8273.44	−7952.17	−7649.26
53	$Z_{(3)}$	1174.93	−9250.09	−8075.16	−7609.40	−7178.79
54	$Z_{(4)}$	1257.56	−9129.84	−7872.27	−7272.76	−6729.25
55	$Z_{(5)}$	1336.69	−9002.02	−7665.32	−6942.72	−6300.34
56	$Z_{(6)}$	1412.12	−8866.99	−7454.87	−6619.71	−5891.69
57	$Z_{(7)}$	1483.67	−8725.12	−7241.45	−6304.11	−5502.90
58	$Z_{(8)}$	1551.18	−8576.79	−7025.61	−5996.29	−5133.54
59	$Z_{(9)}$	1614.50	−8422.41	−6807.90	−5696.55	−4783.14
60	$Z_{(10)}$	–	88080.30	88080.30	72256.53	59503.90
Total					**0**	**−7368.19**

Looking at these results we see that the difference in value using the technical pricing rate of 2% and the market yield of 4% amounts to 7368.19, which represents 7.3% of the maturity benefit. Since the corresponding differences can also become bigger than that consumer protection regulation was introduced in the form of legal quotes. The idea there was that for example 90% of the profit has to allocated to the

M. Koller, *Stochastic Models in Life Insurance*, EAA Series,
DOI: 10.1007/978-3-642-28439-7_12, © Springer-Verlag Berlin Heidelberg 2012

policyholder and the shareholder receives maximally the remaining 10%. Assume that in a concrete year, we have:

- Mathematical reserve: 45'000 USD,
- Investment return: 4.2%,
- Required technical interest rate: 2.0%.

In this case we get the following income statement:

Item	Amount
Investment income	1890 USD
Technical interest	−900 USD
Gross profit	990 USD
of which PH	891 USD
of which SH	99 USD

and we remark the following:

1. The biggest part of the gross profit is allocated to the policyholder and the shareholder receives only the smaller part.
2. Since the return on the assets is random, this is also true for the bonus payment, which however can not become negative.
3. As a consequence the shareholder assumes an additional downside risk if he operates an investment strategy which also allows investment returns below the technical interest rate with a positive probability.
4. It is worth mentioning that the insurance company can also give a higher policyholder bonus than the legal minimum defined by the legal quote. Obviously the shareholder return is then reduced accordingly.

After the split of the gross investment return into policyholder bonus and shareholder return, one needs to look how the bonus is used. Here there are different possibilities. The bonus can be used to

- Reduce the premium. Hence assuming a regular premium of 4900 USD, the policyholder would only need to pay 4001 USD.
- Accumulate it on a bank account. The bonus is invested in a bank account type of investment where a yearly interest rate is credited. At the end of the policy the insured receives the value of this account.
- Increase the benefits. In this case the policyholder bonus is used as a single premium to increase the insurance benefits.

We finally want to revisit the above example with different investment returns:

Return	−5%	0%	2%	5%
Investment income	−2250	0	900	2250
Technical interest	−900	−900	−900	−900
Gross profit	−3150	−900	0	1350
of which PH	0	0	0	1215
of which SH	−3150	−900	0	135

Finally we want to remark that there are two slightly different approaches to legal quotes. In the UK context the insurance company balance sheet is split into policyholder and shareholder funds. The legal quote applies to the policyholder funds. In case of an under-coverage of the policyholder funds, the shareholder is required to compensate for this just as in the example above for an investment return of -5%. In continental Europe, such as in Germany and France the legal quote is applied to the whole balance sheet. The difference between the two concepts is that in the UK context the shareholder receives the full return on his assets. On the other hand the shareholder has to share the return on "his" assets with the policyholder according to the legal quote requirements. In a next step we will look at insurance policies from a different point of view, namely from the policyholders' point of view. Also here we want to apply a market consistent approach and hence the value of the premium before and after paying them to the insurance company is equal, but it is allocated to different stakeholders. We want to illustrate this concept based on an example. We assume that

1. the policy is defined as above as an endowment policy with the parameters set of example 11.4.10, and that
2. the pricing interest rate amounts to $i = 2\%$ with a market interest rate of $i = 4\%$.
3. Furthermore we assume that there is a legal quote of 90% of the difference in present values as shown above.
4. Finally we assume that the present value of the administration charges amounts to 0.2% of the present value of the benefits and that the tax rate of the insurance company amounts to 35%.

In this case we know that:

1. The premium amounts to $P = 1.02 \times 9562.20 = 9753.44$.
2. The present value of the premiums amounts (at 4%) to -1.02×58033.16.
3. The present value of death benefits amounts (at 4%) to 7285.59.
4. The present value of maturity benefits amounts (at 4%) to 28481.29.
5. The present value of surrender payments amounts (at 4%) to 14533.06.
6. The present value of gross profit amounts (at 4%) to 7368.19.

Based on this information we can now decompose the present value of the payments of the policyholder (-59k USD) into its parts:

Type	Stakeholder	Amount	%-age
Mortality	PH	7285.59	12.3%
Maturity	PH	28481.29	48.1%
Surrender	PH	14897.09	25.1%
PH bonus	PH	6631.37	11.2%
Subtotal PH	PH	57295.34	**96.7%**
Admin	Employees of insurer	1160.66	1.9%
Tax	Tax authorities	257.88	0.4%
Profit	SH	478.93	**0.8%**
Total	(equals PV prem.)	59193.82	100.0%

After having seen why policyholder bonus is generated, we need to better understand what is done with the bonus allocated to a policy, since this materially impacts the underlying investment strategies and the corresponding risks. There are from a high level point of view two different ways how gross bonus can be used, namely for the benefit of an individual policyholder or for the entire collective of policyholders together. In the first case the individual policyholder benefits from the excess funds allocated to his policy, be in the form of a reduction of premiums or be it that his benefits are increased. Besides the direct allocation of the gross bonus to individual policyholders, there is also the possibility that a part of the money is used for the entire insurance portfolio together, in the sense of mutualisation of insurance risk. This is known as reserve strengthening and is a consequence that the mathematical reserves for an insurance portfolio are estimated based on statistical methods, which involve some uncertainty. In case a new estimation of the expected reserves needed turns out to be higher than the reserves within the balance sheet, it is necessary to strengthen them in order to be able to honour the corresponding commitments. Now there are two possibilities how such a reserve strengthening is financed, by the shareholder, or as mentioned before by the collective of policyholders by using a part of the gross profits stemming from the in-force portfolio.

12.2 Portfolio Calculations

When doing ALM it is normally important to use efficient calculation processes since a lot of simulations are needed. Let's look at the moment at an insurance portfolio with 1 million polices. During the year end calculation, one needs normally to calculate the mathematical reserves for the past, the current and the next year, in order to save these values in the data base to be able to interpolate them for a possible policy surrender. Assume for the moment that your tariff engine is performing such a calculation in 0.01 s. Hence the whole year end calculation takes you 8 h 20 min. When doing ALM this is obviously too long when requiring 10000 simulations, since this would result in about 3500 years run-time on the same infrastructure. Some acceleration can be gained by using a grid, but even then it seems to make sense to look for faster methods to do the above task. In this section we will look at such approaches, which can concretely be implemented. If we consider a simple set up we have a set of policies $\mathcal{P}$, and each policy can be characterised by its Markov representation, eg the state space S_i, the discount factors (which is normally the same for all polices), the transition probabilities (which are normally structurally similar per different state space) and finally the benefits vectors $a_{ij}(t)$ and $a_i(t)$. In a lot of cases one can also restrict the state space to a common one, which we call S. In order to be concrete we want to have a look at the set of all insurance policies on one life, be it a lump sum or an annuity. In this case we choose $S = \{*, \dagger, \ddagger\}$ as corresponding state space for a person with age ξ, where $\ddagger$ represents the state of a surrendered policy. In this set up we introduce the linear vector space of all insurance policies for this person ξ by

$$\mathcal{F}_\xi = \{x_\xi = (a_{ij}(t), a_i(t)) : i, j \in S \text{ and } t \in \mathbb{N}\}.$$

We now remark that both the mathematical reserve and the expected cash flow operators

$$\Phi_{t,j}(x_\xi) : \mathcal{F}_\sim \to \mathbb{R}, x_\xi \mapsto V_j(t)[x_\xi]$$
$$\Psi_{t,j}(x_\xi) : \mathcal{F}_\sim \to \mathbb{R}, x_\xi \mapsto \mathbb{E}[CF(s)[x_\xi] | X_{t_0} = j]$$

are linear (continuous) functionals from $\mathcal{F}_\xi \to \mathbb{R}$, where ξ denotes the policy considered, t and s the respective times and $j \in S$ a state. When recognising this fact we can now construct the space of all insurance policies for a given portfolio $\mathcal{P} = \{\xi_1, \xi_2, \ldots \xi_n\}$ by defining the respective Cartesian product such as:

$$S = \prod_{i \in \mathcal{P}} S_i, \text{ and}$$

$$\mathcal{F} = \prod_{i \in \mathcal{P}} \mathcal{F}_i.$$

In the same sense we can now define the mathematical reserve and expected cash flow operator of the whole insurance portfolio as the corresponding sum:

$$\Phi_t(x) : \mathcal{F} \to \mathbb{R}, x((\xi_i)_{i \in \mathcal{P}}) \mapsto \sum_{i \in \mathcal{P}} V_{j(\xi_i)}(t)[x_{\xi_i}] \text{ and}$$

$$\Psi_t(x) : \mathcal{F} \to \mathbb{R}, x((\xi_i)_{i \in \mathcal{P}}) \mapsto \sum_{i \in \mathcal{P}} \mathbb{E}[CF(s)[x_{\xi_i}] | X_{t_0} = j(\xi_i)].$$

Until now we have gained nothing except for a more complex representation of what we already know. The way we can now make the whole thing much more efficient is to use the given structure and the fact that the two operators defined above are linear. In the concrete set up where each policy is characterised by the three states above we can define a new "pseudo" state space

$$\tilde{S} = \{x0, x1, \ldots, x120, y0, y1, \ldots, \dagger, \ddagger\},$$

where the states $x0, \ldots, x120$ stand for males which are alive and have the respective age at time t_0. Similarly $y0, \ldots, y120$ stand for the respective females. There is only a need to map the respective $x_\xi \in \mathcal{F}_\xi$ into $\tilde{\mathcal{F}}$. Assume that $\mathcal{R}50 \subset \mathcal{P}$ is a homogeneous set of policies representing the males ages 50. In this case we have the following benefit functions:

$$a_{x50}^{Pre} = \sum_{\xi \in \mathcal{R}50} a_*^{Pre}(\xi)$$

$$a_{x50,x50}^{Post} = \sum_{\xi \in \mathcal{R}50} a_{*,*}^{Post}(\xi)$$

$$a_{x50,\dagger}^{Post} = \sum_{\xi \in \mathcal{R}50} a_{*,\dagger}^{Post}(\xi)$$

$$a_{x50,\ddagger}^{Post} = \sum_{\xi \in \mathcal{R}50} a_{*,\ddagger}^{Post}(\xi)$$

Please note that in order to implement the approach described above, one needs to be careful with respect to the definition of time. In the usual Markov model the time t is calculated with respect to the age of each policyholder in a certain year. For the above purpose, it is useful to enumerate by the number of years into the future, starting at $t = 0$. Hence one needs to correspondingly adjust the time of the benefit functions. We remark that for all the other states in $\tilde{S} \setminus \{\dagger, \ddagger\}$ the same approach is used. It is also worth noting that this "pseudo" Markov chains can be interpreted as "normal" Markov chains, where the initial state of the portfolio is given by a probability distribution, over which one integrates. After having done so, we can now perform a lot of the calculations much faster. After having applied the above mapping into $\tilde{\mathcal{F}}$, the calculations do not depend anymore on the actual size of the portfolio. By this we gain considerable amounts of times when doing the actual calculations. Looking at the sample portfolio above we would possibly use 1h calculation time to perform the mapping on $\tilde{\mathcal{F}}$, including the data base queries. Once this is done the calculation of the mathematical value and expected cash flow operator take some 1 to 2 s on a common laptop. Hence doing ALM in a lot of cases result in run-times of several minutes. In the same sense stress scenarios can be calculated much faster. Finally I would like to mention that this approach has been applied concretely for the examples in Sect. 12.3 using 402 states for $\tilde{\mathcal{F}}$ by also splitting annuities from capital insurance. Using this approach one can also map deferred widows pension using a collective approach and hence one can cover the vast majority of traditional life insurance covers sold by a life insurance company. Since structurally the method is a slight variation of the Markov recursion, this method can actually be implemented using the same core Markov calculation objects.

12.3 Portfolio Dynamics and ALM

The aim of this section is to look at the portfolio dynamics induced by the relationship between assets and liabilities and the corresponding asset liability management (ALM). To this end we fix $(\Omega, \mathcal{A}, P)$, together with a filtration $\mathcal{G} = (\mathcal{G}_t)_{t \in \mathbb{N}}$. We assume that we have represented the benefits and premiums in a suitable ("pseudo") Markov model with state space $\tilde{S}$ and benefits vector space $\tilde{F}$. In this case we can represent the benefits for the entire portfolio by $x = ((a_{i,j}^{Post}(t))_{i,j \in \tilde{S}}, (a_i^{Pre}(t))_{i \in \tilde{S}})_{t \in \mathbb{N}} \in \tilde{F}$, and we know that we can decompose this into

$$x = x^{Benefits} + x^{Premium},$$

where $x^{Benefits}$ and $x^{Premium}$ represent the corresponding benefit vectors in $\tilde{\mathcal{F}}$ for benefits and premiums, respectively. At this point the bonus concept enters. If for example the benefits are increased, the corresponding $x^{Benefits}$ is increased accordingly. Formally one introduces hence a random vector α_t which represents the relative benefit level. This means that $\alpha_0 = 1$ and also that α is previsible. For traditional bonus promises, the bonus allocated to the individual policy become a guarantee, means that α is increasing in t for each trajectory and hence we define the new benefit representation of the policy at time t as follows:

$$\hat{x}^{Benefits} = ((\alpha_t \times a_{i,j}^{Post}(t))_{i,j \in \tilde{s}}, (\alpha_t \times a_i^{Pre}(t))_{i \in \tilde{s}})_{t \in \mathbb{N}},$$

remarking that this is now a random quantity, since α is a random vector. In consequence the entire portfolio after bonus allocation is given by

$$\hat{x} = \hat{x}^{Benefits} + x^{Premium}.$$

The mechanism to increase α is performed by using the allocated bonus as a single premium. In a next step we want have a look at a concrete example. The first step is to define a stochastic model, which generates the corresponding states of the world. In the concrete set up we are assuming a world with a constant interest rate with one risky asset (say a share) with has a non constant, stochastic volatility. We are using the Heston model, which is given by

$$dV_t = \kappa(\theta - V_t)dt + \eta\sqrt{V_t}dW_t^1$$
$$dS_t = \mu S_t \, dt + \sqrt{V_t} \, S_t \, dW_t^2.$$

This model is given by two stochastic differential equations, where the first equation describes the volatility of the shares. This volatility process has a structural similarity with the interest rate models we have seen before, in the sense that the volatility process is also mean reverting. In order to solve this stochastic differential equation system we can use a numerical method such as the Milstein scheme (see for example [KP92]). After having done this, it is important to understand how the simulation works. In principle one does first a loop over the different simulation and calculates the quantities, which need to be analysed. The following code performs the corresponding task:

```
1  # 4. Stepper
2  # --------------------------
3  for i in 1... n:
4      sim.vNewTrajectory()
5      (pl, cf, ... ) = CalcPV(sim, lvp, lvl)
   ... keep and analyse results for run i
```

Hence first one generates a new trajectory of the world (line 4, in this case based on the Heston model), and secondly on calculates the development of the insurance portfolio, as outlined above in the subroutine `CalcPV` (line 5). Hence it makes sense to look at this part of the code in order to understand what happens:

```
...  code left away
 1  for i in 1... MaxTime:
 2       TargetSurplus   = ... target surplus level required
 3       perf = ((1+iRF) * (1-EquityProportion) +
             EquityProportion* sim.dGetValue(1,float(i)) / sim.dGetValue(1,float(i-1.)))
 4       cf[i] = psi[i-1] * lvl.dGetCF(i) + lvp.dGetCF(i)
 5       l[i] = psi[i-1] * lvl.dGetDKTilde(i) + lvp.dGetDKTilde(i) - cf[i]
 6       a[i] = a[i-1] * perf - cf[i]
 7       ExAssets = a[i] - l[i]
 8       # Calculate Excess Assets and Pl, afterwards adjust assets by pl for SH
 9       if ExAssets > 0.:
10           ExAssets = max(0., ExAssets - TargetSurplus)
11           Bonus = ExAssets * RelBonusAllocation * LegalQuote
12           pl[i-1] = ExAssets * RelBonusAllocation * (1-LegalQuote)
13       else:
14           Bonus = 0.
15           pl[i-1] = ExAssets
16       a[i] -= pl[i-1]
17       e[i] += pl[i-1]
18       psi[i] = psi[i-1] + Bonus / (psi[i-1] * (lvl.dGetDKTilde(i) - lvl.dGetCF(i)))
```

What this code does, is the following:

1. In a first step, the minimal required surplus of the policyholder funds (line 1) and the performance of the assets (line 2) is calculated. We see that in the concrete set up we have a mixture of assets. A part of them yielding risk free and the reminder, the equity portion having an equity yield.
2. In lines 4, 5 and 6 the cash-flows and the assets and liabilities are calculated at the end of the period. We see that

$$\Psi_i(\hat{x}^{Benefits}) = \texttt{psi[i-1]} \ * \ \texttt{lvl.dGetCF(i)}, \text{and}$$
$$\Psi_i(x^{Premium}) = \texttt{lvp.dGetCF(i)}.$$

3. In a next (lines 7 to 15) step the excess assets are calculated in order to determine, whether there is a bonus in the corresponding year. If there are excess assets (line 9), the bonus is calculated.
4. Finally the benefit level for the subsequent year is calculated (line 18).

We remark that the initialisation of the code plus the analysis have been left in order to focus on the essential parts of the calculation. After having done this, we want to have a look at some sample output of the program. Figures 12.1 and 12.2 show the mathematical reserves and the expected cash flows for the benefits and the premiums respectively. Moreover Fig. 12.3 show the quantiles of the profit and loss over time for the 5, 10, 33, 50, 67, 90 and 95% quantiles for both the profit and loss account and the corresponding dividends. We remark that this figure shows that the underlying portfolio is insufficiently financed and suffers considerable losses between the time 5 and 15, and also that the losses start earlier in cases where we observe an adverse equity performance. The output of the program looks as follows:

```
Input Parameter / Main Results
-------------------------------
Simulations                                           5000
>>    PV Benefits ..... .....DK 1   =   22,786,998,031
>>    PV Premium. ..... .....DK 0   =   -4,091,249,171
>>    Mathematical Reserves..DK     =   18,695,748,860
>>    Underlying Assets .....AO     =   22,635,000,000
>>    Shareholder Equity.....EO     =    1,584,450,000
Distribution of Economic Profit
-------------------------------
Min Return                                          0.0962
Max Return                                          0.2473
Average                                             0.1792
 0.5%  Quantile                                     0.1288
 1.0%  Quantile                                     0.1324
 5.0%  Quantile                                     0.1452
10.0%  Quantile                                     0.1529
25.0%  Quantile                                     0.1654
99.5%  Quantile                                     0.2289
Time used
---------
Step                Calc EP Time used     2.3515 s
Step                Calc PF Time used     3.6497 s
Step                Calc PF Time used     3.2491 s
Step              AO and EO Time used     0.0190 s
Step    Simulation Stepper Time used    81.8312 s
Time used for preparation                 9.2 s
Time used for simulations                81.8 s
Time per simulation                       0.0163 s
```

We see in particular that the implementation via the "pseudo" Markov model results in an extremely fast calculation of the portfolio in about 9 seconds. Also the simulation is performed very fast, despite the fact that a relatively slow laptop was used. The output of the above run is summarised also in Fig. 12.4.

12.4 Variable Annuities (VA)

12.4.1 Introduction

In this section we will document and analyse the methods used for the valuation and hedging of the VA portfolio. From a high level point of view the concepts of arbitrage free pricing theory together with Itô-Calculus is used to immunise and hedge

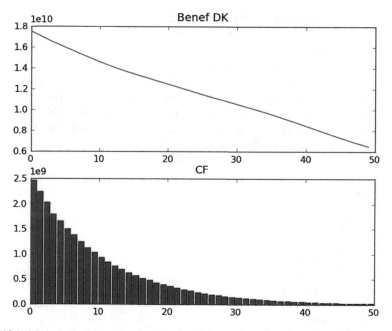

Fig. 12.1 Mathematical reserve and expected cash flows of portfolio (benefits only)

the liabilities outstanding as good as possible with a dynamic hedging strategy. More concretely the aim is to represent the values of the liabilities in terms of a Taylor approximation with respect to the respective relevant market variables such as equity returns, interest rate movement and volatility of the funds. The corresponding first and second order partial derivatives of the liability values (and also assets) are called "Greeks". When using assets with the same underlying value and Greeks means that the change in economic equity moves in parallel for the respective approximations, thereby reducing the risk up to higher order errors of both assets and liabilities. The concept of dynamic hedging foresees the updating of the underlying liability values and Greeks and the corresponding re-balancing of the assets in continuous time in order to result in an optimal hedge. In reality the recalculation of the liability values and the re-balancing is done on a less frequent basis, resulting in a tracking or hedging error.

For simple forms of liabilities the underlying calculations can be performed by closed form formulas. In the concrete set up, the liabilities are more complex and hence valuation and calculation of the Greeks are done via simulation ("Monte Carlo") and depend on the Economic Scenario Generator used. The expected values are estimated in the simulation context by suitable averages and the partial derivatives are estimated with a difference quotient with a suitable small shock.

There are several types of performance guarantees for unit linked policies and one may often choose them a la carte, with higher risk charges for guarantees that are riskier for the insurance company. The first type is comprised of guaranteed

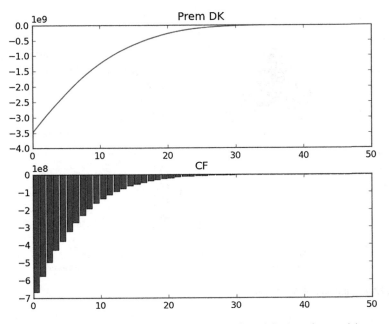

Fig. 12.2 Mathematical reserve and expected cash flows of portfolio (premiums only)

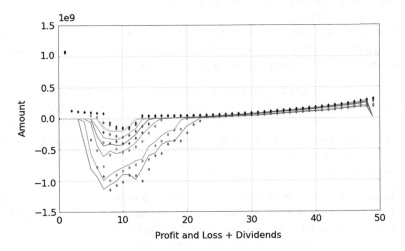

Fig. 12.3 Quantiles for profits and losses

minimum death benefits (GMDB), which can be received only if the owner of the contract dies.

GMDBs come in various forms, in order of increasing risk to the insurance company:

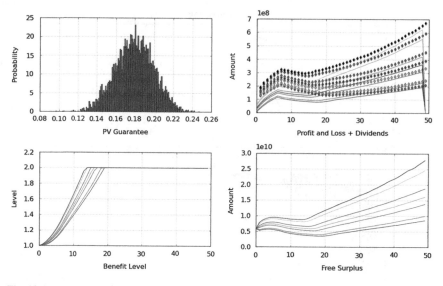

Fig. 12.4 Summary of ALM analysis

- Return of premium (a guarantee that you will not have a negative return),
- Roll-up of premium at a particular rate (a guarantee that you will achieve a minimum rate of return, greater than 0),
- Maximum anniversary value (looks back at account value on the anniversaries, and guarantees you will get at least as much as the highest values upon death),
- Greater of maximum anniversary value or particular roll-up.

Unlike death benefits, which the contract holder generally can't time, living benefits pose significant risk for insurance companies as contract holders will likely exercise these benefits when they are worth the most. Annuities with guaranteed living benefits (GLBs) tend to have high fees commensurate with the additional risks underwritten by the issuing insurer.

Some GLB examples, in no particular order:

- Guaranteed Minimum Income Benefit (GMIB, a guarantee that one will get a minimum income stream upon annuitisation at a particular point in the future)
- Guaranteed Minimum Accumulation Benefit (GMAB, a guarantee that the account value will be for a certain amount at a certain point in the future)
- Guaranteed Minimum Withdrawal Benefit (GMWB, a guarantee similar to the income benefit, but one that doesn't require annuitising)
- Guaranteed-for-life Income Benefit (a guarantee similar to a withdrawal benefit, where withdrawals begin and continue until cash value becomes zero, withdrawals stop when cash value is zero and then annuitisation occurs on the guaranteed benefit amount for a payment amount that is not determined until annuitisation date.)

In order to value this guarantee, one needs to rely on option pricing techniques such as the Black-Scholes formula. The price for a put-option with payout $C(T, P) = \max(K - S; 0)$ at time t and strike price K and equity price S is given by:

$$P = K \times e^{-r \times T} \times \Phi(-d_2) - S_0 \times \Phi(-d_1),$$

$$d_1 = \frac{\log(S_0/K) + (r + \sigma^2/2) \times T}{\sigma \times \sqrt{T}},$$

$$d_2 = d_1 - \sigma \times \sqrt{T},$$

$$\Phi(x) = \int_{-\infty}^{x} \frac{1}{\sqrt{2\pi}} \exp(\frac{-\zeta^2}{2}) d\zeta.$$

The reader should be reminded that the formula is based on the efficient market hypothesis which requests that the following holds:

- Deep and friction-less market, and
- Absence of arbitrage.

In order to understand how these options are synthetically "constructed" one needs to understand the concept of a replicating portfolio. Hence one holds at every point in time a portfolio P_t with the aim that this portfolio matches at time T just the payout of the option mentioned above. In order to construct such portfolios one usually uses the "greeks". These greek letters represent the sensitivity of an option in case of a change to the underlying economic parameters such as equity price, interest rate levels, etc. We have the following relationships:

$$\Delta_P = \frac{\partial P}{\partial S}$$
$$= \Phi(d_1),$$
$$\Gamma = \frac{\partial^2 P}{\partial S^2}$$
$$= \frac{\Phi'(d_1)}{S \times \sigma \times \sqrt{T}},$$
$$\nu = \frac{\partial P}{\partial \sigma}$$
$$= S \times \Phi'(d_1) \times \sqrt{T - t},$$
$$\rho = \frac{\partial P}{\partial r}$$
$$= -(T - t) \times K \times e^{-r \times (T-t)} \times \Phi(-d_2).$$

All these partial derivatives serve to approximate the change in value of the underlying derivatives when the fundamental parameters change. To do this one uses Itô-Formula for a given SDE such as

$$dX_t = \mu_t \, dt + \sigma_t \, dB_t,$$

follows:

$$df(t, X_t) = \left(\frac{\partial f}{\partial t} + \mu_t \frac{\partial f}{\partial x} + \frac{\sigma_t^2}{2} \frac{\partial^2 f}{\partial x^2} \right) dt + \sigma_t \frac{\partial f}{\partial x} dB_t,$$

where we clearly see the various greeks. In the concrete context the respective approximations looks as follows:

$$\Delta f(S, t, r, \sigma) = \underbrace{\frac{\partial f}{\partial S}}_{Delta} \Delta S + \frac{1}{2} \underbrace{\frac{\partial^2 f}{\partial S^2}}_{Gamma} (\Delta S)^2 + \underbrace{\frac{\partial f}{\partial t}}_{Theta} \Delta t$$

$$+ \underbrace{\frac{\partial f}{\partial r}}_{Rho} \Delta r + \underbrace{\frac{\partial f}{\partial \sigma}}_{Vega} \Delta \sigma + \dots$$

Note that this is the based for the hedging program on one side and allows to also determine the hedging error as a consequence of the higher order terms on the other.

Based on the above partial derivatives, it is now possible to define different hedging strategies, one of them being a "delta-hedge". The idea is to define at a point in time t a portfolio P_t consisting of cash and shares, which have the same value and for which the partial derivative with respect to equity price S is the same. Hence we look for a Taylor approximation of order 1 in the variable S. Figure 12.5 shows such a delta hedge.

12.4.2 Product Design

Over all the products need to be understood in the context that there are three phases (see also Fig. 12.6):

1. Accumulation phase without the possibility of withdrawal,
2. The phase of minimal withdrawal during which the policyholder still has the ability to invest in risky assets, and
3. The deccumulation phase where a minimum—constant annuity is granted until the person dies.

Before explaining the various variants in detail, it makes sense to have a closer look at Fig. 12.6. In the concrete set up we are looking at a "virtual" product where the policyholder buys a VA at age 60 and invests 100'000 USD in a 100% equity funds. For the first 5 years (until age 65) there are no additions nor withdrawals to this fund and the value changes as a consequence of the market movement. At age 65 the person starts to withdraw (according to the contract) 5% of the maximum fund value

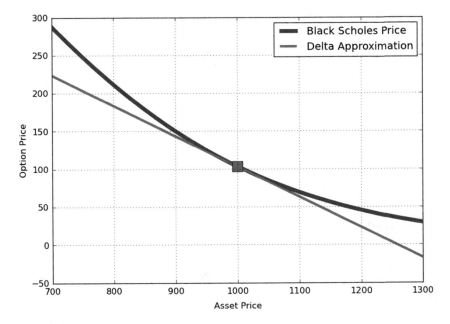

Fig. 12.5 δ–Hedging

at that time and the original amount. Hence, depending on the past funds performance he has the right to withdraw no more than the minimally guaranteed 5'000 USD. Depending on the fund performance there is now a risk for the insurance company since the funds, still investing into equities might run out of money and the company has to further pay the respective monies. This phase ends when the insurance enters the income phase (which starts in the example at age 85). At that time, the funds is derisked and treated as a fixed amount income annuity, where the minimal amount paid to the policyholder is determined just as in the previous phase. Again there is a risk that the fund at age 85 is insufficient to bear the payouts thereafter.

For illustratory purposes we have assumed in the example that the insured person lives until the age of 100, which is obviously high. In reality the random life span of the policyholder makes this product still more complex since mutualisation between policyholders enters. For now it is however clear that the risk for the insurance company is an increased life span and hence a longer consumption period.

There are two types of complexity intrinsic to these VA products, namely the many possible options which can be granted and sold to the policyholder and hence the understanding of the different features. On the other hand it becomes obvious from Fig. 12.6 that market developments which could lead to losses for the insurance company if nothing is done. The technique to mitigate these losses and to replicate these guarantees is to set up a corresponding hedging programme.

GMDB. The Guaranteed minimal death benefit is to some degree the easiest option to each VA. The aim of this benefit is to ensure a minimal death benefit in

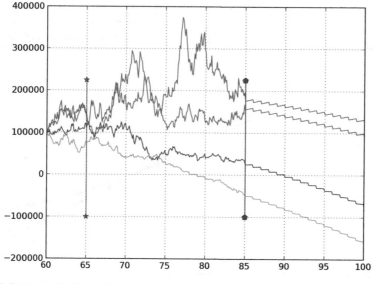

Fig. 12.6 Phases of a VA product

case the policyholder dies. Assume, for example, that the policyholder has invested
100'000 USD at age 60 and dies during the first phase at age 63. We assume also
that the fund value has dropped to 68'000 USD. The easiest GMDB is to guarantee a
minimal fixed amount in case of death, say 120'000 USD for arguments sake. In this
case the loss to the insurance company is 52'000 USD ($= 120000 - 68000$). From a
financial point of view the insurance company has hence written a put option at age
63 with a strike price of 120'000 USD. In consequence one can consider a GMDB
portfolio as put option portfolio, where the strike prices correspond to the respective
death benefits at a certain time and where the amount of put is weighted according
to the expected number of people dying at this time.

In some instances the definition of the death benefit is bit more involved. As
explained below, but from a principal point of view, it follows the scheme outlined
above. In the following we denote with $D(x)$ the total death benefit in case a person
dies at age x.

Basic death benefit: This minimal amount is insured with every VA policy. It is
defined as the maximum of current funds value $F(x)$ and the difference between the
sum of the paid premium $P(x)$ and the sum of the past withdrawals $W(x)$. Hence as
a formula:

$$D^{\text{basic}}(x) = \max(F(x), \sum_{k<x}\{P(k) - W(k)\}).$$

Total death benefit: On top of this basic death benefit the policyholder can choose
between two different options, for which he pays a premium. In both cases the total
death benefit equals the maximum of basic death benefit and the additional amount,

hence

$$D^{\text{total}}(x) = \max(D^{\text{basic}}(x), D_i(x)),$$

where $D_i(x)$ denotes the death benefit according to option i.

Option 1: $D_1(x)$ In this case the idea is that there is a death benefit which relates to a ratchet (high water mark) of the funds value for a certain period, plus all the premiums paid after the end of the ratchet period. The ratchet is determined quarterly, eg with respect to times $\mathbb{T} := \{t_1, t_2, \ldots t_n\}$. Hence we have:

$$D_1(x) = \max\{F(t_k) : t_k \in \mathbb{T}, t_k < x\} + \sum_{\max(t : t \in \mathbb{T}) < k < x} (P(k) - W(k)).$$

Option 2: $D_2(x)$ Here the situation is very similar to above with exception that the premiums are compounded with a fixed interest rate i and that also the paid out withdrawals are subtracted. This then results in the following formula:

$$D_2(x) = \max\{0, \max\{F(t_k) : t_k \in \mathbb{T}, t_k < x\}$$
$$+ \sum_{\max(t : t \in \mathbb{T}) < k < x} (P(k) - W(k) \times (1+i)^{(x-k)}\}.$$

Please note that the compounding (in terms of time) is limited to a certain age x; hence afterwards the respective amounts do not increase anymore, but the GMDB remain in place. From this point of view the GMDB is "whole life".

GMWB. Considering a GMWB written for a single premium, this process involves the random movement of the funds. Figure 12.6 shows this, assuming that the insured person dies at age 100. In a next step the trajectories with losses need to be weighted with the corresponding people alive and not having surrendered at a certain point in time. Figure 12.7 shows this with some 10 trajectories.

After having looked at this sample trajectories one needs to look at the expected cash flow profile stemming from the guarantee. As before we start with a 60 year old person with a deferral period of 5 years (hence starting to withdraw at age 65 at 5% p.a.) with a income phase starting at age 85. We furthermore assume a single premium of USD 100'000. Figure 12.8 shows this. The red curve illustrates the expected losses and compares them with a guaranteed (fixed) annuity where the money is invested risk free. Two things become obvious:

- The guarantee starts to bite in most cases after some 25 years when the people are 85+ years old.
- The percentage of the guarantee (on the lower part of the figure) increases with increasing age and exceeds 20% at age 80.
- In average this guarantee equals about 14% of the funds and has a rather long duration (of 26).

From a formal point of view the following quantities are important:

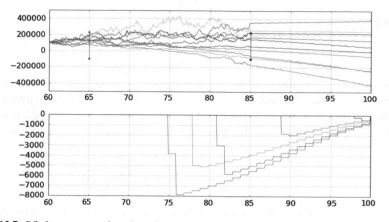

Fig. 12.7 P/L for some sample trajectories

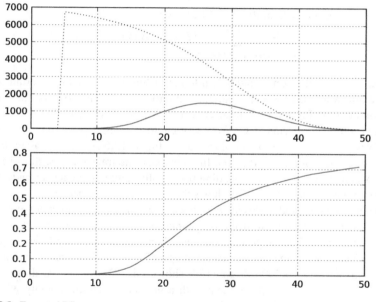

Fig. 12.8 Expected P/L

Funds Value: $F(t)$ denotes the funds value, **GAWA:** $GAWA(t)$ is the maximal amount which can be consumed under the GMWB policy, **GWA:** $GWA(t)$ is the maximal amount which can be withdrawn per year without needing a partial surrender.

It is worth noting that both quantities $GAWA$ and GWA depend on the time and (potentially) also on the funds performance, in the sense that an additional bonus could be allocated to the funds. Each time a withdrawal is made, the $GAWA$ is reduced by this amount. Depending on the contract specificities, the guarantee consists of being able to continue withdrawing (until a certain point in time or until death) even when the maximal amount is exceeded.

The following aspects are worth noting:

- The initial $GAWA$ and GWA are a function of the funds value at the first time the policyholder decides to withdraw money. This particularly means that policyholder behaviour plays an important role.
- In some instances the $GAWA$ has also a ratchet feature, which could mean that the $GAWA$ is locked in at each quarter end and can not fall thereafter (before withdrawing). Normally the GWA is then expressed as a fixed percentage of the $GAWA$.

Formulae:

In the following section I summarise the economics of this sort of contract. Assume the following:

- Person aged x_0 purchases such a GMWB and pays a single premium EE;
- Assume that the person starts to withdraw at age sw and that the income phase starts at age s;
- We use the following notation:

  - $F(t)$: Funds value at time t. To be more precise we denote with $F(t)^-$ and $F(t)^+$ the value of the funds before and after withdrawal of the annuity, respectively. In consequence we have $F(t)^+ = F(t)^- - R(t)$.
  - $GWB(t)$: GWB ("Guaranteed Withdrawal Balance") value at time t. To be more precise we denote with $GWB(t)^-$ and $GWB(t)^+$ the value of the funds before and after withdrawal of the annuity, respectively. In consequence we have $GWB(t)^+ = GWB(t)^- - R(t)$.
  - $f(x)$: GAWA percentage if person starts to withdraw at age x;
  - $GAWA(t)$: maximal allowable withdrawal benefit (usually $= GWB \times f(x)$).
  - $R(t)$: actual amount withdrawn. Note that we have the following: $R(\xi) = 0$ for $\xi < sw$, $0 \leq R(\xi) \leq GAWA(x)$, for all $\xi \in [sw, s[$, and $R(\xi) = GAWA(x)$, for all $\xi \geq s$, assuming that the "for life option" is in place and identifying x and t in the obvious way, eg $x(t) = t - t_0 + x_0$

- With $\eta(t, \tau) \in \mathbb{R}$, we denote the fund performance during the time interval $[t, \tau]$, with $\tau > t$.
- We do not allow for changes in funds and lapses at this time and also do not consider the death of the person insured. This would add some complexity, where actually the annuities need to be weighted with the respective probabilities $_t p_x$ and in the same sense the respective death cover weighted with $_t p_x q_x$. For the moment assume that $\sum \tau \geq 0$ stands for "until death".
- By $X(t)$ we denote the loss at time t occurring from GMWB guarantees. It is obvious that under these premises the value of the total guarantee $Y = \sum_{\tau \geq 0} (1 + r(\tau))^{-\tau} X(\tau)$, where $r(\tau)$ represents the risk free interest between $[0, \tau]$.

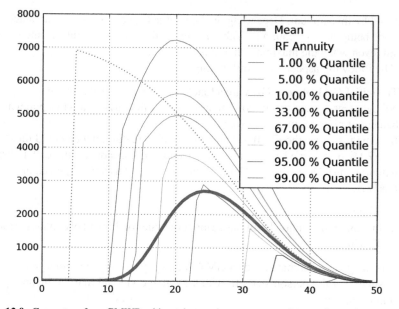

Fig. 12.9 Guarantees for a GMWB with ratchet; x: time; y: payout of respective option

As described above we assume for example the following:

$$f(x) = \begin{cases} 5\% & \text{if } x \in [55, 74], \\ 6\% & \text{if } x \in [75, 84], \\ 7\% & \text{if } x \geq 85. \end{cases}$$

For the recursion we have at time $t_0 = 0$

$$F(0) = EE,$$
$$GWB(0) = EE,$$
$$GAWA(0) = f(sw) \times GWB(0).$$

Afterwards from time $t - 1 \rightsquigarrow t$ we have the following, assuming that the withdrawals takes place before the end of the first quarter, once a year.

$$F(t)^- = (1 + \eta(t - 1, t)) \times F(t - 1)^+,$$
$$GWB(t)^- = \max\{GWB(t - 1)^+, \max_{k=0,1,\dots,4}\{1 + \eta(t - 1, t - 1 + \frac{k}{4})\}$$
$$\times F(t - 1)^+\},$$
$$GAWA(t) = \max(GAWA(t - 1), f(sw) \times GWB(t)^-),$$
$$F(t)^+ = \max(0, F(t)^- - R(t)),$$

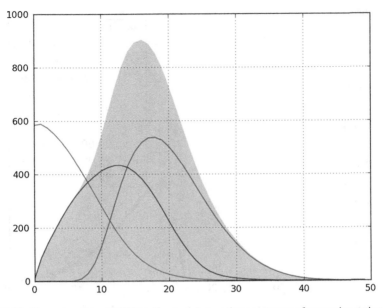

Fig. 12.10 Guarantees for a GMWB without ratchet; x: time; y: payout of respective option

$$GWB(t)^+ = \max(0, GWB(t)^- - R(t)),$$
$$X(t) = \max(0, R(t) - F(t)^-),$$
$$\pi(Y) = E^Q \left[\sum_{\tau \geq 0} (1 + r(\tau))^{-\tau} X(\tau) \right].$$

If we now pick a given, mortality cover—the simplest one—namely the payment of $GWB(t)$ in case of death, we can calculate the value of the insurance option as follows:

$$\pi(Y) = E^Q \left[\sum_{\tau \geq 0}^{\infty} (1 + r(\tau))^{-\tau} X(\tau) \times {}_\tau p_{x_0} \right],$$

where we assume that no guarantee fee is charged. The inclusion of a guarantee fee can be viewed as an additional annuity which consumes the funds, but only until its depletion.

Next we look at a concrete example, once with and once without ratchet guarantee (aka "step-up"). Figures 12.9 and 12.10 show the way the guarantee manifests in the case with and without the ratchet option. The Fig. 12.9 plots the quantiles (with respect to a risk neutral simulation) of the guarantees X. Moreover Fig. 12.10 shows the additional guarantee fee (green) and the GMDB payment (blue). Note that all

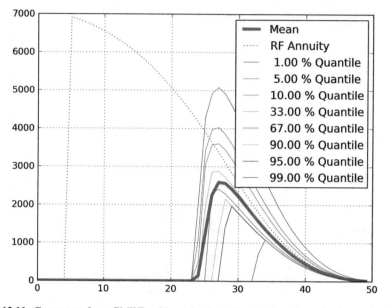

Fig. 12.11 Guarantees for a GMWB with ratchet including the funds transfer feature; x: time; y: payout of respective option

guarantee payments occur in later years as a consequence that the ratchet does not affect the funds value, but rather the (virtual) GWB account. Only when the funds run out of money the guarantee starts to bite.

Appendix A
Notes on Stochastic Integration

This appendix summarises definitions and results in the area of stochastic integration and martingales. For obvious reasons we will not present proofs for all results. In fact we will only give a survey of various results and then refer to the corresponding literature. Foremost we refer to the monographs [28] and [48].

A.1 Stochastic Processes and Martingales

Definition A.1.1 A probability space $(\Omega, \mathcal{A}, P)$ is called *filtered*, if there exists a family of σ-algebras $F = (\mathcal{F}_t)_{t \geq 0}$ such that

1. $\mathcal{F}_0 \supset \{A \in \mathcal{A} | P(A) = 0\}$,
2. $\mathcal{F}_s \subset \mathcal{F}_t$ for $s \leq t$.

The filtration is called right continuous, if $\mathcal{F}_t = \bigcap_{t' < t} \mathcal{F}_{t'}, \forall t \geq 0$.

Definition A.1.2 A random variable $T : \Omega \to [0, \infty]$ is a *stopping time*, if $\{T \leq t\} \in \mathcal{F}_t$ for all $t \in \mathbb{R}_+$.

Theorem A.1.3 *T is a stopping time if and only if $\{T < t\} \in \mathcal{F}_t$ for all $t \in \mathbb{R}_+$.* ([48] *Theorem* 1.1.1.)

Definition A.1.4 Let X, Y be stochastic processes. X is called *modification of* Y, if

$$X_t = Y_t \quad P-\text{almost surely for all t.}$$

X and Y are *indistinguishable*, if

$$X_t = Y_t, \forall t \quad P-\text{almost surely.}$$

M. Koller, *Stochastic Models in Life Insurance*, EAA Series,
DOI: 10.1007/978-3-642-28439-7, © Springer-Verlag Berlin Heidelberg 2012

Definition A.1.5 1. A stochastic process is called *càdlàg* (continue à droite, limites à gauche), if its trajectories are right continuous with left limits.
2. A stochastic process is called *càglàd* (continue à gauche, limites à droite), if its trajectories are left continuous with right limits.
3. A stochastic process is *adapted*, if $X_t \in \mathcal{F}_t$ (X_t is $\mathcal{F}_t$-measurable).

Theorem A.1.6 1. *Let Λ be an open set and X be an adapted càdlàg process. Then $T := \inf\{t \in \mathbb{R}_+ : X_t \in \Lambda\}$ is a stopping time.*
2. *Let S, T be stopping times and $\alpha > 1$. Then the following random variables are also stopping times*: $\min(S, T), \max(S, T), S + T, \alpha \cdot T$.

Proof [48] Theorem 1.1.3 and 1.1.5.

Definition A.1.7 Let $(\Omega, \mathcal{A}, (\mathcal{F}_t)_{t \geq 0}, P)$ be a filtered probability space. A stochastic process X is called *martingale*, if

- $X_t \in L^1(\Omega, \mathcal{A}, P)$, i.e. $E[|X_t|] < \infty$,
- $E[X_t|\mathcal{F}_s] = X_s$ holds for $s < t$.

Remark A.1.8 If one replaces in the previous equation "$=$" by "$\leq$" ("$\geq$"), the process X is called supermartingale (submartingale).

Theorem A.1.9 *Let X be a supermartingale. Then the following conditions are equivalent*:

1. *The mapping $T \to \mathbb{R}, t \mapsto E[X_t]$ is right continuous.*
2. *There exists a unique modification Y of X which is càdlàg.*

Proof [48] Theorem 1.2.9.

Theorem A.1.10 *Let X be a martingale. Then there exists a unique càdlàg modification Y of X.*

Theorem A.1.11 (Doob's stopping theorem) *Let X be a right continuous martingale with closure X_∞, i.e. $X_t = E[X_\infty|\mathcal{F}_t]$. Moreover, let S and T be stopping times such that $S \leq T$ P-almost surely. Then the following statements hold*:

1. $X_S, X_T \in L^1(\Omega, \mathcal{A}, P)$,
2. $X_S = E[X_T|\mathcal{F}_S]$.

Proof [48] Theorem 1.2.16.

Definition A.1.12 Let X be a stochastic process and T be a stopping time. The stopped stochastic process $(X_t^T)_{t \geq 0}$ is defined by $X_t^T = X_{\min(t,T)}$ for all $t \geq 0$.

Theorem A.1.13 (Jensen's inequality) *Let $\phi : \mathbb{R} \to \mathbb{R}$ be a convex function and $X \in L^1(\Omega, \mathcal{A}, P)$ with $\phi(X) \in L^1(\Omega, \mathcal{A}, P)$. Furthermore let $\mathcal{G}$ be a σ-algebra. Then the following inequality holds:*

$$\phi \circ E[X|\mathcal{G}] \leq E[\phi(X)|\mathcal{G}].$$

Proof [48] Theorem 1.2.19.

A.2 Stochastic Integrals

In this section we present a short introduction to the theory of stochastic integration. We follow the approach of [48].

Essentially one can understand a stochastic integral with respect to a semi-martingale as a pathwise Stieltjes integral, and the latter should be known from lectures in analysis. The main idea for this type of integrals is to consider the limit of sums of the form

$$\sum f(T_k)(T_{k+1} - T_k)$$

for partitions with decreasing mesh size. In the following we assume that a filtered probability space $(\Omega, \mathcal{A}, (\mathcal{F}_t)_{t \geq 0}, P)$, which satisfies the common regularity conditions, is given.

Definition A.2.1 1. A stochatic process H is called *simple predictable*, if it can be represented as

$$H_t = H_0 \cdot \chi_{\{0\}}(t) + \sum_{i=1}^{n} H_i \cdot \chi_{]T_i, T_{i+1}]}(t),$$

where

$$0 = T_1 \leq \ldots \leq T_{n+1} < \infty$$

is a finite family of stopping times and the $H_i \in \mathcal{F}_t, (H_i)_{i=0,\ldots,n}$ are finite P-almost everywhere.

The set of simple predictable processes will be denoted by $\mathbb{S}$. Furthermore, $\mathbb{S}_u$ denotes the set $\mathbb{S}$ equipped with the topology of uniformly convergence in (t, ω) on $\mathbb{R} \times L^\infty(\Omega, \mathcal{A}, P)$.
2. The vector space of finite, real valued random variables equipped with the convergence in probability is denoted by $\mathbb{L}^0$.

Next, we want to define the expression $\int H\, dX$ for certain processes $(X_t)_{t \in \mathbb{R}}$ and $(H_t)_{t \in \mathbb{R}}$. In order to be able to call such an operator, denoted by I_X, integral, it should at least be linear and a theorem like the Lebesgue convergence theorem should hold.

For the convergence theorem we assume the following continuity: If H^n converges uniformly to H, then $I_X(H^n)$ should converge in probability to $I_X(H)$.

Given a process X. We define $I_X : \mathbb{S} \to \mathbb{L}^0$ by

$$I_X(H) = H_0 X_0 + \sum_{i=1}^{n} H_i \left(X^{T_i} - X^{T_{i+1}} \right),$$

where

$$H_t = H_0 \cdot \chi_{\{0\}}(t) + \sum_{i=1}^{n} H_i \cdot \chi_{]T_i, T_{i+1}]}(t).$$

This definition of $I_X(H)$ is independent of the representation of H.

Definition A.2.2 (*Total semimartingale*) A stochastic process $(X_t)_{t \geq 0}$ is called *total semimartingale*, if

1. X is càdlàg and
2. I_X is a continuous mapping from $\mathbb{S}_u$ to $\mathbb{L}^0$.

Definition A.2.3 (*Semimartingale*) A stochastic process $(X_t)_{t \geq 0}$ is a *semimartingale*, if X^t (compare with Definition A.1.12) is a total semimartingale for all $t \in [0, \infty[$.

Remark A.2.4 Thus semimartingles are defined as well behaving integrators.

The following theorem summarizes the most important properties of the operator I_X:

Theorem A.2.5 1. *The set of all semimartingales is a vector space.*
2. *Let Q be a measure which is absolutely continuous with respect to P. Then every P-semimartingale is also a Q-semimartingale.*
3. *Let $(P_n)_{n \in \mathbb{N}}$ be a sequence of probability measures such that $(X_t)_{t \geq 0}$ is a P_n-semimartingale for each n. Then $(X_t)_{t \geq 0}$ is an R-semimartingale, for $R = \sum_{n \in \mathbb{N}} \lambda_n P_n$, where $\sum_{n \in \mathbb{N}} \lambda_n = 1$.*
4. *(Stricker's Theorem) Let X be a semimartingale with respect to the filtration $(\mathcal{F}_t)_{t \geq 0}$, and $(\mathcal{G}_t)_{t \geq 0}$ be a sub-filtration of $(\mathcal{F}_t)_{t \geq 0}$ such that X is adapted to $(\mathcal{G}_t)_{t \geq 0}$. Then X is a $\mathcal{G}$-semimartingale.*

Proof The statements follow from the definition of a semimartingale. The proofs, which are recommended as an exercise to the reader, can be found in [48] Chap. II.2.

Now we want to characterise the class of semimartingales.

Theorem A.2.6 *Every adapted process with càdlàg paths and finite variation on compact sets is a semimartingale.*

Proof This theorem is based on the fact that

$$|I_X(H)| \leq ||H||_u \int_0^\infty |dX_s|,$$

where $\int_0^\infty |dX_s|$ denotes the total variation.

Theorem A.2.7 *Every square integrable martingale with càdlàg paths is a semimartingale.*

Proof Let X be a square integrable martingale with $X_0 = 0, H \in \mathbb{S}$. The continuity of the operators I_X is a consequence of the following inequality:

$$
\begin{aligned}
E[(I_X(H))^2] &= E\left[\left(\sum_{i=0}^n H_i(X^{T_i} - X^{T_{i+1}})\right)^2\right] \\
&= E\left[\sum_{i=0}^n H_i^2(X^{T_i} - X^{T_{i+1}})^2\right] \\
&\leq ||H||_u^2 E\left[\sum_{i=0}^n (X^{T_i} - X^{T_{i+1}})^2\right] \\
&= ||H||_u^2 E\left[\sum_{i=0}^n \left(X^{T_i\,2} - X^{T_{i+1}\,2}\right)\right] \\
&= ||H||_u^2 E\left[X_{T^{n+1}}{}^2\right] \\
&\leq ||H||_u^2 E\left[X_{T^\infty}{}^2\right].
\end{aligned}
$$

Example A.2.8 Brownian motion is a semimartingale.

Now, after defining semimartingales, we want to enlarge the class of integrands. A class which is very well suited for this purpose is the set of càglàd processes. We will use this class, since for it the proofs remain relatively simple.

Definition A.2.9 The set of adapted càdlàg (càglàd) processes is denoted by $\mathbb{D}$ ($\mathbb{L}$, respectively). Further, $b\mathbb{L}$ denotes the set of processes $X \in \mathbb{L}$ with bounded paths.

Up to now we are familiar with the topology of the uniform convergence (on $\mathbb{S}_u$) and the topology of the convergence in probability on $\mathbb{L}^0$. We define a further notion of convergence.

Definition A.2.10 Let $t \geq 0$ and H be a stochastic process. Then we set

$$H_t^* = \sup_{0 \leq s \leq t} |H_s|.$$

A sequence $(H^n)_{n \in \mathbb{N}}$ converges uniformly on compact sets in probability (short: convergence in ucp-topology) to H, if

$$(H^n - H)_t^* \to 0$$

in probability for $n \to \infty$ and all $t \geq 0$.

$\mathbb{D}_{ucp}, \mathbb{L}_{ucp}$ and $\mathbb{S}_{ucp}$ denote the corresponding sets equipped with the ucp-topology.

Remark A.2.11 1. The ucp-topology is metrizable. An equivalent metric is for example:

$$d(X, Y) = \sum_{i=1}^{\infty} \frac{1}{2^n} E\left[\min(1, (X - Y)_n^*)\right].$$

2. $\mathbb{D}_{ucp}$ is a complete metric space.

The following theorem is essential for the extension of the integral I_X.

Theorem A.2.12 *The vectorspace $\mathbb{S}$ is dense in $\mathbb{L}$ with respect to the ucp-topology.*

Proof [48] Theorem 2.4.10.

Note that, I_X can be extended if we can show that I_X is continuous. To show the continuity we start with a definition.

Definition A.2.13 Let $H \in \mathbb{S}$ and X be a semimartingale. Then we define $J_X : \mathbb{S} \to \mathbb{D}$ by

$$J_X(H) = H_0 X_0 + \sum_{i=0}^{n} H_i \left(X^{T_i} - X^{T_{i+1}}\right),$$

where

$$H_t = H_0 \cdot \chi_{\{0\}}(t) + \sum_{i=1}^{n} H_i \cdot \chi_{]T_i, T_{i+1}]}(t)$$

for $H_i \in \mathcal{F}_{T_i}$ and stopping times $0 = T_1 \leq \ldots \leq T_{n+1} < \infty$.

Definition A.2.14 (*Stochastic integral*) Let $H \in \mathbb{S}$ and X be a càdlàg process. Then $J_X(H)$ is called the stochastic integral of H with respect to X and we use the notations:

$$H \cdot X := \int H_s \, dX_s := J_X(H).$$

Now we have defined the stochastic integral on $\mathbb{S}$, and we want to extend it onto $\mathbb{L}$. For the extension we need the following theorem.

Theorem A.2.15 *Let X be a semimartingale. Then the mapping $J_X : \mathbb{S}_{ucp} \to \mathbb{D}_{ucp}$ is continuous. Also the extension of J_X onto $\mathbb{S}_{ucp}$ will be called stochastic integral, and the notations of Definition A.2.14 will be used also for the extension.*

Proof [48] Theorem 2.4.11.

Remark A.2.16 To extend J_X onto $\mathbb{D}$ we use the fact that $\mathbb{D}_{ucp}$ is a *complete* metric space.

The process $J_X(H) = \int H_s\, dX_s$ evaluated at the time $t \geq 0$ will be denoted by

$$H \cdot X_t := \int_0^t H_s\, dX_s := \int_{[0,t]} H_s\, dX_s.$$

A.3 Properties of the Stochastic Integral

After defining the stochastic integral we will now summarise its properties. We will concentrate on the statements without giving proofs.

Theorem A.3.1 1. *Let T be a stopping time. Then* $(H \cdot X)^T = H \cdot \chi_{[0,T]}$.
   $X = H \cdot X^T$.
2. *Let $G, H \in \mathbb{L}$ and X be a semimartingale. Then also $Y := H \cdot X$ is a semartingale. Furthermore, we have*

$$G \cdot Y = G \cdot (H \cdot X) = (G \cdot H) \cdot X.$$

Proof [48] Theorem 2.5.12 and 2.5.19.

Definition A.3.2 Let X be a càdlàg process. Then we define

$$X_-(t) = \lim_{s \uparrow t} X(s),$$
$$\Delta X(t) = X(t) - X_-(t).$$

Definition A.3.3 A random partition σ of $\mathbb{R}$ is a finite sequence of stopping times such that

$$0 = T_0 \leq T_1 \leq \ldots \leq T_n < \infty.$$

A sequence $(\sigma_n)_{n \in \mathbb{N}}$ of random partitions of $\mathbb{R}$ converges to the identity, if the following conditions hold:

1. $\lim_{n \to \infty} \left(\sup_k T_k^n \right) = \infty$ P-almost surely,
2. $\|\sigma_n\| := \sup_k |T_{k+1}^n - T_k^n|$ converges P-almost surely to 0.

Let Y be a process and σ be a random partition. Then we define

$$Y^\sigma := Y_0 \cdot \chi_{\{0\}} + \sum_k Y_{T_k} \cdot \chi_{]T_k, T_{k+1}]}.$$

Remark A.3.4 It is easy to show that

$$\int Y_s^\sigma dX_s = Y_0 X_0 + \sum_k Y_{T_k} \left(X^{T_{k+1}} - X^{T_k} \right)$$

for all semimartingales X and all Y in $\mathbb{S}, \mathbb{D}$ and $\mathbb{L}$.

Using random partitions one calculate the stochastic integral by the following theorem.

Theorem A.3.5 *Let X be a semimartingale, $Y \in \mathbb{D}$ and $(\sigma_n)_{n \in \mathbb{N}}$ be a sequence of random partitions which converges to the identity. Then*

$$\int_{0^+} Y_s^{\sigma_n} dX_s = \sum_k Y_{T_k^n} \left(X^{T_{k+1}^n} - X^{T_k^n} \right)$$

converges in ucp-topology towards the stochastic integral $\int (Y_-) dX$.

Proof [48] Theorem 2.5.21.

Definition A.3.6 Let X and Y be semimartingales. Then we define

$$[X, X] = ([X, X]_t)_{t \geq 0} \text{ the quadratic variation process by}$$
$$[X, X] := X^2 - 2 \int X_- dX,$$

and accordingly

$$[X, Y] := XY - \int X_- dY - \int Y_- dX$$

is the *covariation process*.

Theorem A.3.7 *Let X be a semimartingale. Then the following statements hold*:

1. $[X, X]$ *is càdlàg, monotone increasing and adapted.*
2. $[X, X]_0 = X_0^2$ *and* $\Delta[X, X] = (\Delta X)^2$.
3. *If a sequence $(\sigma_n)_{n \in \mathbb{N}}$ of random partitions converges to 1, then*

$$X_0^2 + \sum_i (X^{T_{i+1}^n} - X^{T_i^n})^2 \longrightarrow [X, X] \text{ in ucp-topology for n} \to \infty.$$

4. *Let T be a stopping time. Then* $[X^T, X] = [X, X^T] = [X^T, X^T] = [X, X]^T$.

Proof [48] Theorem 2.6.22.

Remark A.3.8 • The mapping $(X, Y) \mapsto [X, Y]$ is bilinear and symmetric.
• The polarization identity holds:

$$[X, Y] = \frac{1}{2}([X + Y, X + Y] - [X, X] - [Y, Y]).$$

Theorem A.3.9 *The bracket process $[X, Y]$ of two semimartingales X and Y has paths of bounded variation on compact sets and it is a semimartingale.*

Proof [48] Corollary 2.6.1.

Theorem A.3.10 (Partial integration)

$$d(XY) = X_- dY + Y_- dX + d[X, Y].$$

Proof [48] Corollary 2.6.2.

Theorem A.3.11 *Let M be a local martigale. Then the first and second of the following statements are equivalent and the third is a consequence of the first two*

1. M is martingale with $E[M_t^2] \leq \infty \ \forall t \geq 0$,
2. $E\big[[M, M]_t\big] < \infty \ \forall t \geq 0$,
3. $E[M_t^2] = E\big[[M, M]_t\big] \forall t \geq 0$.

Proof [48] Corollary 2.6.4.

Theorem A.3.12 *Let X, Y be semimartingales and $H, K \in \mathbb{L}$. Then the following statements hold:*

1. $[H \cdot X, K \cdot Y]_t = \int_0^t H_s K_s d[X, Y]_s \ \forall \, t \geq 0$,
2. $[H \cdot X, H \cdot X]_t = \int_0^t H_s^2 d[X, X]_s \ \forall \, t \geq 0$.

Proof [48] Theorem 2.6.29.

Theorem A.3.13 (Itô-formula) *Let X be a semimartingale and $f \in C^2(\mathbb{R})$. Then*

$$f(X_t) - f(X_0) = \int_{0^+}^t f'(X_s^-)dX_s + \frac{1}{2} \int_{0^+}^t f''(X_s^-)d[X, X]_s^{cont}$$

$$+ \sum_{0 < s \leq t} \left\{ f(X_s) - f(X_s^-) - f'(X_s^-)\Delta X_s \right\}.$$

Proof [48] Theorem 2.7.32.

Remark A.3.14 A function $f \in C^2(\mathbb{R})$ has the deterministic integral representation

$$f(t) - f(0) = \int_0^t f'(s)ds.$$

For a stochastic integral two further terms appear. The term

$$\frac{1}{2} \int_{0^+}^t f''(X_s^-)d[X, X]_s^{cont}$$

is due to the quadratic variation of the process and the term

$$\sum_{0<s\leq t} \{f(X_s) - f(X_s^-) - f'(X_s^-)\Delta X_s\}$$

is due to the jumps of the process.

Theorem A.3.15 (Transformation theorem) *Let V be a stochastic process with right continuous paths of bounded variation. Furthermore, let $f \in C^1(\mathbb{R})$. Then $(f(V_t))_{t\geq 0}$ is a process of bounded variation and*

$$f(V_t) - f(V_0) = \int_{0+}^{t} f'(V_{s-})dV_s + \sum_{0<s\leq t} (f(V_s) - f(V_{s-}) - f'(V_{s-})\Delta V_s).$$

Theorem A.3.16 (Itô-formula) *Let X be a continuous semimartingale and $f \in C^2(\mathbb{R})$. Then also $f(X)$ is a semimartingale and it satisfies*

$$f(X_t) - f(X_0) = \int_{0+}^{t} f'(X_s)dX_s + \frac{1}{2}\int_{0+}^{t} f''(X_s)d[X,X]_s.$$

Appendix B
Examples

This book is complemented by set of example files which can be downloaded from the webpage http://www.xxx.xxx/???.

This appendix explains how to install these files.

B.2 Description

1. Excel-files with the corresponding DLL for the calculation of the discrete Markov model. These files require Excel 97 or later.
2. Stand-alone program for the calculation of the discrete models. This requires Windows 95, Windows NT 3.5 or later.

B.1.1 Directory C:/markov

File	Size in byte	Description
invalev2.xls	852,480	disability insurance
kt1995.xls	141,312	example of a mortality table
markov.xls	175,616	general Markov model
mathserv.dll	150,016	Markov calculator for Excel files
regsvr32.exe	27,648	program for the registration
rmark.exe	350,208	Markov model for input files
tdf2bsp.xls	359,424	general life insurance

M. Koller, *Stochastic Models in Life Insurance*, EAA Series,
DOI: 10.1007/978-3-642-28439-7, © Springer-Verlag Berlin Heidelberg 2012

B.1.2 Directory C:/markov/pension

Example files for rmark.exe: pension.

File	Size in byte	Description
d40.txt	1,611	discount function (4%)
evk90f.txt	16,096	$p_{ij}(x)$ for female
evk90m.txt	16,128	$p_{ij}(x)$ for male
ln_null.txt	10	payments in arrear
lvarevk1.txt	2,222	payments in advance

B.1.3 Directory C:/markov/endowment

Example files for rmark.exe: endowment policy.

File	Size in byte	Description
d35.txt	1,637	discount function (3.5%)
gkm95.txt	3,864	$p_{ij}(x)$ for male
ln_G.txt	839	payments in arrear
lv_null.txt	41	payments in advance

B.1.4 Directory C:/markov/disability

Example files for rmark.exe: disability insurance.

File	Size in byte	Description
d35.txt	1,637	discount function (3.5%)
iunda95.txt	86,506	$p_{ij}(x)$ for male
ln_ia4.txt	839	payments in arrear
lv_ia4.txt	41	payments in advance

B.2 Installation

The first step is to download and run the program *markov.exe*. This extracts the archive containing the example files.

Excel files: One has to register the DLL which will be required for the calculations. For this open a DOS-box ("cmd.exe") and change the current directory to `c:\markov` by using the commands

```
c:
cd c:\markov
```

then the DLL is registered by the command

```
regsvr32 mathserv.dll
```

Now a message window should appear, which states that the registration was successful. Afterward one can use the following three Excel files. Please note, that for the calculations it is required that the macro security settings of Excel allow the execution of macros.

markov.xls	For an arbitrary model. Programmable with maximal 15 states.
tod2bsp.xls	For a life insurance on one life with stochastic interest rate (Sect. 6.5)
inval.xls	For a disability insurance. (Sect. 6.6 and Example 10.3.2)

rmark.exe: The program *rmark* can be used directly. It requires four input files, which define the model. The results are written into an output file, whose name has to be entered.

The required input files are the following:

pre.dat	payments in advance a_i^{Pre}
post.dat	payments in arrear a_{ij}^{Post}
pij.dat	transition probabilities p_{ij}
discount.dat	discount rate v_t

The file pre.dat has the following structure:

```
15   1   3   700
```

This means, that at the age 15 in state 1 a payment in advance of 700 USD is due. The successive state (here 3) is not used. Trivial payments are not required explicitly.

To chose the pre-file, one uses the button *new pre file* and selects as file type "all files *.*". Then the file can be selected.

The post-file and the pij-file (transition probabilities) have the same structure. The entry shown above would mean, that at the age 15 a payment of 700 is due, if one moves from state 1 to state 3.

The discount-file contains the yearly discount rates. The syntax of this file is different to the one above.

```
15   0.96
```

This means, that at the age 15 a discount rate of 0.96 is applied.

After creating these files one can view and edit them by using the edit-button.

To start the calculations also the following additional informations have to be entered in the corresponding fields.

```
initial time:       e.g. 120
final time:         e.g. 65
number of states:   e.g. 3 (Note: the program is limited
to 15 states)
new output file:    e.g. by the button 'new output file'
```

Examples of input files can be found in the subdirectories.

Appendix C
Mortality Rates in Germany

The following mortality tables are based on [12]. They differentiate between the mortality of smokers ("S") non smokers ("NoS"). In order to keep the appendix at a reasonable size, we only included fundamentals of 2nd order.

	DAV2008T q_x^{NoS}	DAV2008T q_x^{S}	DAV2008T q_y^{NoS}	DAV2008T q_y^{S}	DAV2008P i_x	DAV2008P i_y
0	0.003797	0.004562	0.004562	0.003797	–	–
1	0.000289	0.000316	0.000316	0.000289	–	–
2	0.000237	0.000256	0.000256	0.000237	–	–
3	0.000190	0.000205	0.000205	0.000190	–	–
4	0.000151	0.000164	0.000164	0.000151	–	–
5	0.000122	0.000136	0.000136	0.000122	–	–
6	0.000100	0.000116	0.000116	0.000100	–	–
7	0.000086	0.000104	0.000104	0.000086	–	–
8	0.000078	0.000096	0.000098	0.000078	–	–
9	0.000074	0.000093	0.000096	0.000075	–	–
10	0.000076	0.000096	0.000101	0.000079	–	–
11	0.000083	0.000107	0.000114	0.000088	–	–
12	0.000095	0.000129	0.000139	0.000103	–	–
13	0.000114	0.000166	0.000182	0.000128	–	–
14	0.000140	0.000226	0.000250	0.000163	–	–
15	0.000168	0.000309	0.000346	0.000203	–	–
16	0.000197	0.000410	0.000461	0.000245	–	–
17	0.000222	0.000518	0.000585	0.000284	–	–
18	0.000229	0.000617	0.000696	0.000300	–	–
19	0.000229	0.000689	0.000776	0.000306	–	–
20	0.000226	0.000730	0.000818	0.000305	–	–

M. Koller, *Stochastic Models in Life Insurance*, EAA Series, DOI: 10.1007/978-3-642-28439-7, © Springer-Verlag Berlin Heidelberg 2012

	DAV2008T q_x^{NoS}	DAV2008T q_x^{S}	DAV2008T q_y^{NoS}	DAV2008T q_y^{S}	DAV2008P i_x	DAV2008P i_y
21	0.000220	0.000737	0.000821	0.000299	–	–
22	0.000213	0.000724	0.000802	0.000291	–	–
23	0.000207	0.000694	0.000770	0.000283	–	–
24	0.000202	0.000654	0.000732	0.000280	–	–
25	0.000198	0.000610	0.000698	0.000281	–	–
26	0.000195	0.000568	0.000675	0.000287	–	–
27	0.000193	0.000533	0.000666	0.000298	–	–
28	0.000193	0.000508	0.000675	0.000316	–	–
29	0.000193	0.000491	0.000699	0.000338	–	–
30	0.000195	0.000482	0.000737	0.000366	–	–
31	0.000200	0.000479	0.000783	0.000401	–	–
32	0.000211	0.000481	0.000835	0.000446	–	–
33	0.000228	0.000487	0.000892	0.000503	–	–
34	0.000251	0.000498	0.000953	0.000574	–	–
35	0.000281	0.000512	0.001019	0.000657	–	–
36	0.000314	0.000531	0.001095	0.000747	–	–
37	0.000350	0.000557	0.001183	0.000840	–	–
38	0.000390	0.000592	0.001292	0.000940	–	–
39	0.000432	0.000636	0.001426	0.001044	–	–
40	0.000479	0.000691	0.001590	0.001158	0.000085	0.000110
41	0.000532	0.000758	0.001786	0.001284	0.000091	0.000130
42	0.000592	0.000839	0.002021	0.001427	0.000103	0.000154
43	0.000661	0.000935	0.002300	0.001592	0.000120	0.000183
44	0.000740	0.001052	0.002635	0.001781	0.000142	0.000217
45	0.000831	0.001184	0.003013	0.001999	0.000171	0.000255
46	0.000933	0.001332	0.003434	0.002246	0.000205	0.000298
47	0.001045	0.001488	0.003877	0.002521	0.000247	0.000346
48	0.001165	0.001654	0.004345	0.002820	0.000296	0.000400
49	0.001293	0.001831	0.004839	0.003145	0.000353	0.000461
50	0.001427	0.002024	0.005371	0.003493	0.000417	0.000531
51	0.001571	0.002244	0.005973	0.003877	0.000489	0.000610
52	0.001727	0.002491	0.006643	0.004302	0.000571	0.000701
53	0.001892	0.002771	0.007400	0.004765	0.000663	0.000803
54	0.002069	0.003078	0.008228	0.005272	0.000768	0.000918
55	0.002259	0.003418	0.009146	0.005830	0.000889	0.001046
56	0.002469	0.003795	0.010166	0.006458	0.001031	0.001187
57	0.002696	0.004219	0.011315	0.007150	0.001198	0.001342
58	0.002950	0.004689	0.012596	0.007930	0.001394	0.001513

	DAV2008T q_x^{NoS}	DAV2008T q_x^S	DAV2008T q_y^{NoS}	DAV2008T q_y^S	DAV2008P i_x	DAV2008P i_y
59	0.003238	0.005227	0.014060	0.008821	0.001628	0.001704
60	0.003569	0.005845	0.015736	0.009841	0.001907	0.001919
61	0.003957	0.006571	0.017697	0.011032	0.002240	0.0021640
62	0.004426	0.007440	0.020021	0.012460	0.002637	0.002446
63	0.004988	0.008492	0.022809	0.014146	0.003106	0.002775
64	0.005647	0.009777	0.026154	0.016104	0.003657	0.003159
65	0.006429	0.011358	0.030212	0.018386	0.004295	0.003611
66	0.007331	0.013227	0.034906	0.020975	0.005029	0.004150
67	0.008371	0.015395	0.040197	0.023890	0.005866	0.004802
68	0.009545	0.017891	0.046105	0.027090	0.006822	0.005602
69	0.010885	0.020653	0.052397	0.030621	0.007920	0.006601
70	0.012413	0.023679	0.058986	0.034509	0.009195	0.007862
71	0.014154	0.026983	0.065811	0.038755	0.010699	0.009463
72	0.016105	0.030367	0.072335	0.043291	0.012501	0.011492
73	0.018711	0.033738	0.078306	0.049210	0.014687	0.014045
74	0.021204	0.037409	0.084395	0.054366	0.017359	0.017221
75	0.024078	0.041402	0.090588	0.060003	0.020622	0.021121
76	0.027406	0.045866	0.097098	0.066157	0.024580	0.025844
77	0.031289	0.050877	0.104044	0.072965	0.029323	0.031495
78	0.035750	0.056589	0.111650	0.080259	0.034926	0.038187
79	0.040993	0.063039	0.119774	0.088339	0.041448	0.046053
80	0.046993	0.070338	0.128577	0.096948	0.048942	0.055252
81	0.053819	0.078516	0.138032	0.106077	0.057478	0.065971
82	0.061505	0.087533	0.147930	0.115568	0.067174	0.078420
83	0.070057	0.097419	0.158306	0.125262	0.078230	0.092822
84	0.079747	0.108215	0.169141	0.135491	0.090946	0.109382
85	0.090633	0.119895	0.180441	0.146282	0.105705	0.128263
86	0.102906	0.132530	0.192169	0.157652	0.122883	0.149552
87	0.116564	0.146092	0.204529	0.169601	0.142677	0.173229
88	0.131464	0.160372	0.216983	0.181814	0.164845	0.199182
89	0.147608	0.175288	0.229803	0.194252	0.188452	0.227241
90	0.164792	0.190761	0.242648	0.206650	0.211715	0.257305
91	0.182792	0.206673	0.255655	0.219168	0.235254	0.287368
92	0.201453	0.222933	0.268635	0.231470	0.259344	0.317428
93	0.219863	0.239433	0.281813	0.243168	0.283987	0.347485
94	0.237813	0.256125	0.294800	0.253984	0.309182	0.377540
95	0.255996	0.272931	0.307866	0.265468	0.334929	0.407592
96	0.274491	0.289797	0.321095	0.274491	0.361228	0.437641
97	0.292905	0.306997	0.334319	0.292905	0.388078	0.467687

	DAV2008T q_x^{NoS}	DAV2008T q_x^{S}	DAV2008T q_y^{NoS}	DAV2008T q_y^{S}	DAV2008P i_x	DAV2008P i_y
98	0.311537	0.324694	0.348396	0.311537	0.415481	0.497731
99	0.330385	0.342733	0.362612	0.330385	0.443436	0.527772
100	0.349446	0.361062	0.377310	0.349446	0.471943	0.557810
101	0.368713	0.379675	0.392584	0.368713	0.501002	0.587845
102	0.388182	0.398564	0.408529	0.388182	0.530612	0.617878
103	0.407846	0.417759	0.424861	0.407846	0.560775	0.647908
104	0.427697	0.437213	0.442022	0.427697	0.591490	0.677935
105	0.447726	0.456916	0.460114	0.447726	0.622757	0.70796
106	0.467921	0.477006	0.477006	0.467921	0.654576	0.737981
107	0.488271	0.497189	0.497189	0.488271	0.686947	0.768001
108	0.508762	0.517650	0.517650	0.508762	0.719869	0.798017
109	0.529376	0.538377	0.538377	0.529376	0.753344	0.828030
110	0.550097	0.559353	0.559353	0.550097	0.787371	0.858041
111	0.570904	0.580560	0.580560	0.570904	0.821950	0.888050
112	0.591772	0.601975	0.601975	0.591772	0.857081	0.918055
113	0.612677	0.623571	0.623571	0.612677	0.892764	0.948058
114	0.633589	0.645315	0.645315	0.633589	0.928999	0.978058
115	0.654476	0.667170	0.667170	0.654476	0.965786	1.000000
116	0.675302	0.689091	0.689091	0.675302	1.000000	1.000000
117	0.696026	0.711028	0.711028	0.696026	1.000000	1.000000
118	0.716604	0.732920	0.732920	0.716604	1.000000	1.000000
119	0.736988	0.754701	0.754701	0.736988	1.000000	1.000000
120	0.757123	0.776292	0.776292	0.757123	1.000000	1.000000
121	1.000000	1.000000	1.000000	1.000000	1.000000	1.000000

Appendix D
Mortality Rates in Switzerland

The following mortality rates are taken from the Swiss Kollektivtarif1995. They are useful for calculations of concrete examples. The Kollektivtarif 1995 differentiates the fundamentals between those used for insurances which provide a fixed capital benefit (GKM/F) and those used for insurances which provide a pension (GRM/F). The last letter M (F, resp.) indicates that the table is for male (female, resp.).

	GKM95 q_x	GRM95 q_x	GKF95 q_y	GRF95 q_y	GKM95 i_x	GKF95 i_y
15	0.00158	0.00129	0.00030	0.00032	0.00307	0.00136
16	0.00160	0.00129	0.00033	0.00032	0.00316	0.00145
17	0.00160	0.00129	0.00034	0.00032	0.00327	0.00159
18	0.00160	0.00129	0.00034	0.00032	0.00340	0.00179
19	0.00158	0.00129	0.00033	0.00033	0.00354	0.00202
20	0.00155	0.00129	0.00033	0.00033	0.00368	0.00228
21	0.00151	0.00129	0.00034	0.00034	0.00383	0.00254
22	0.00146	0.00130	0.00036	0.00037	0.00398	0.00282
23	0.00142	0.00130	0.00039	0.00040	0.00413	0.00309
24	0.00139	0.00130	0.00042	0.00042	0.00428	0.00335
25	0.00136	0.00130	0.00045	0.00045	0.00442	0.00360
26	0.00133	0.00130	0.00048	0.00048	0.00456	0.00383
27	0.00131	0.00130	0.00051	0.00051	0.00470	0.00405
28	0.00130	0.00130	0.00054	0.00055	0.00483	0.00425
29	0.00130	0.00130	0.00058	0.00058	0.00496	0.00443
30	0.00130	0.00131	0.00061	0.00061	0.00509	0.00459
31	0.00131	0.00131	0.00065	0.00065	0.00521	0.00473
32	0.00133	0.00133	0.00069	0.00069	0.00533	0.00486
33	0.00136	0.00136	0.00073	0.00073	0.00545	0.00497

	GKM95 q_x	GRM95 q_x	GKF95 q_y	GRF95 q_y	GKM95 i_x	GKF95 i_y
34	0.00140	0.00140	0.00077	0.00077	0.00558	0.00507
35	0.00145	0.00144	0.00082	0.00082	0.00571	0.00516
36	0.00150	0.00150	0.00087	0.00087	0.00586	0.00525
37	0.00158	0.00157	0.00092	0.00092	0.00602	0.00534
38	0.00166	0.00165	0.00098	0.00097	0.00620	0.00543
39	0.00176	0.00175	0.00103	0.00103	0.00640	0.00553
40	0.00187	0.00186	0.00109	0.00108	0.00662	0.00565
41	0.00200	0.00199	0.00114	0.00114	0.00689	0.00577
42	0.00214	0.00213	0.00119	0.00119	0.00719	0.00592
43	0.00231	0.00230	0.00124	0.00124	0.00753	0.00610
44	0.00250	0.00249	0.00129	0.00129	0.00793	0.00630
45	0.00271	0.00270	0.00135	0.00135	0.00839	0.00653
46	0.00295	0.00294	0.00142	0.00142	0.00891	0.00680
47	0.00323	0.00322	0.00150	0.00149	0.00951	0.00710
48	0.00355	0.00353	0.00160	0.00157	0.01020	0.00744
49	0.00391	0.00387	0.00172	0.00165	0.01097	0.00782
50	0.00431	0.00422	0.00187	0.00174	0.01185	0.00823
51	0.00476	0.00458	0.00205	0.00184	0.01284	0.00869
52	0.00527	0.00496	0.00226	0.00195	0.01395	0.00918
53	0.00583	0.00536	0.00251	0.00206	0.01519	0.00970
54	0.00645	0.00580	0.00277	0.00219	0.01658	0.01026
55	0.00713	0.00627	0.00305	0.00234	0.01812	0.01083
56	0.00788	0.00679	0.00335	0.00250	0.01983	0.01143
57	0.00869	0.00735	0.00366	0.00268	0.02171	0.01204
58	0.00957	0.00796	0.00397	0.00288	0.02379	0.01265
59	0.01052	0.00864	0.00427	0.00310	0.02607	0.01326
60	0.01155	0.00937	0.00458	0.00334	0.02856	0.01384
61	0.01266	0.01018	0.00487	0.00359	0.03129	0.01440
62	0.01384	0.01107	0.00514	0.00384	0.03427	0.01490
63	0.01511	0.01196	0.00551	0.00413	0.03750	0.00000
64	0.01646	0.01282	0.00609	0.00447	0.04102	0.00000
65	0.01807	0.01370	0.00689	0.00491	0.04482	0.00000
66	0.02003	0.01464	0.00791	0.00545	0.00000	0.00000
67	0.02234	0.01569	0.00915	0.00608	0.00000	0.00000
68	0.02500	0.01689	0.01062	0.00680	0.00000	0.00000
69	0.02801	0.01828	0.01233	0.00757	0.00000	0.00000
70	0.03137	0.01989	0.01428	0.00840	0.00000	0.00000
71	0.03508	0.02175	0.01647	0.00926	0.00000	0.00000
72	0.03914	0.02389	0.01892	0.01015	0.00000	0.00000

	GKM95 q_x	GRM95 q_x	GKF95 q_y	GRF95 q_y	GKM95 i_x	GKF95 i_y
73	0.04355	0.02629	0.02161	0.01105	0.00000	0.00000
74	0.04831	0.02891	0.02457	0.01196	0.00000	0.00000
75	0.05342	0.03175	0.02779	0.01288	0.00000	0.00000
76	0.05887	0.03476	0.03127	0.01387	0.00000	0.00000
77	0.06468	0.03793	0.03503	0.01500	0.00000	0.00000
78	0.07084	0.04123	0.03907	0.01635	0.00000	0.00000
79	0.07735	0.04465	0.04339	0.01797	0.00000	0.00000
80	0.08421	0.04816	0.04800	0.01992	0.00000	0.00000
81	0.09141	0.05174	0.05290	0.02224	0.00000	0.00000
82	0.09897	0.05538	0.05810	0.02498	0.00000	0.00000
83	0.10688	0.05906	0.06360	0.02815	0.00000	0.00000
84	0.11513	0.06276	0.06940	0.03171	0.00000	0.00000
85	0.12374	0.06647	0.07552	0.03559	0.00000	0.00000
86	0.13269	0.07018	0.08195	0.03972	0.00000	0.00000
87	0.14200	0.07389	0.08870	0.04404	0.00000	0.00000
88	0.15166	0.07779	0.09578	0.04850	0.00000	0.00000
89	0.16166	0.08206	0.10319	0.05305	0.00000	0.00000
90	0.17202	0.08688	0.11093	0.05763	0.00000	0.00000
91	0.18272	0.09242	0.11902	0.06221	0.00000	0.00000
92	0.19378	0.09883	0.12745	0.06680	0.00000	0.00000
93	0.20518	0.10625	0.13622	0.07163	0.00000	0.00000
94	0.21693	0.11450	0.14535	0.07693	0.00000	0.00000
95	0.22904	0.12328	0.15484	0.08295	0.00000	0.00000
96	0.24149	0.13258	0.16470	0.08970	0.00000	0.00000
97	0.25429	0.14241	0.17492	0.09703	0.00000	0.00000
98	0.26745	0.15277	0.18552	0.10494	0.00000	0.00000
99	0.28095	0.16366	0.19649	0.11344	0.00000	0.00000
100	0.29480	0.17507	0.20785	0.12253	0.00000	0.00000

Appendix E
Java Code for the Calculation
of the Markov Model

The following code is an example of a (non optimised) implementation of the
Markov recursion in Java. Please note, that we removed some consistency checks
from the original code in order to increase the readability of this printed version.

```java
/*
 * MarkovClass.java
 */

public class MarkovClass {
        double  dPij[][];      /* Transition probability [Age][Index]   */
        double  dPost[][];     /* Postnumerando benefits [Age][Index]   */
        double  dPre[][];      /* Prenumerando benefits  [Age][State]   */
        double  dDisc[][];     /* Discount Factor        [Age][State]   */
        int     iFrom[];       /* [Index]       */
        int     iTo[];         /* [Index]       */
        int     iMat2Idx[][];  /* [From][To]    */
        double  dV[][];        /* [Age][States] */
        double  dVTemp[];      /* [States]      */

        int   iMaxTimes;
        int   iMaxIndex;
        int   iMaxStates;
        int   iStates;
        int   iStart;
        int   iStop;
        int   iAdvancedDisc;
        int   iT;
        int   iNrIndex = 0;
        int   iC1, iC2;
        int   iErr;

        boolean bDKCalculated = false;
        boolean bGetData = false;

        /** Creates a new instance of MarkovClass */
        /***************************************/
```

```java
public MarkovClass(int iMaxStateInput, int iMaxTime) {

    /* Define Max allocatable memory */

    iMaxTimes  = iMaxTime;
    iMaxStates = iMaxStateInput;
    iMaxIndex  = iMaxStates * iMaxStates;
    dPij       = new double[iMaxTimes][iMaxIndex];    /* [Age][Index] */
    dPost      = new double[iMaxTimes][iMaxIndex];    /* [Age][Index] */
    dPre       = new double[iMaxTimes][iMaxStates];   /* [Age][State] */
    dDisc      = new double[iMaxTimes][iMaxStates];   /* [Age][State] */

    iFrom      = new int[iMaxIndex];                  /* [Index] */
    iTo        = new int[iMaxIndex];                  /* [Index] */
    iMat2Idx   = new int[iMaxStates][iMaxStates];;    /* [From][To] */
    dV         = new double[iMaxTimes][iMaxStates];   /* [State] */
    dVTemp     = new double[iStates];                 /* [State] */

    for(iC1=0; iC1 < iMaxStates; ++ iC1)
    {
        for(iC2=0; iC2 < iMaxStates; ++ iC2)
        {
        iMat2Idx[iC1][iC2] = -1;
        }
    }

bDKCalculated = false;
bGetData = false;

}
/** Member functions */
/** 1. vReset */
public void   vReset(){
bDKCalculated = false;}

/** 2. vSetStartTime */
public void         vSetStartTime(int iTime){
iStart = iTime;
bDKCalculated = false;}

/** 3. vSetStopTime */
public void         vSetStopTime(int iTime){
iStop = iTime;
bDKCalculated = false;}

/** 4. vSetNrStates */
public void         vSetNrStates(int iNrStatesIpt){
iStates = iNrStatesIpt;
bDKCalculated = false;}

/** 5. vSetGetData */
public void         vSetGetData(boolean bStatus){
bGetData = bStatus;}

/** 6. vSetPre (Sets Prenumerando Cash Flows) */
public double       dSetPre(int iTimeInput, int iFromInput,
                                int iToInput, double dValue){
    if (bGetData == true)
    {
      return(dPre[iTimeInput][iFromInput]);
    }
    bDKCalculated = false;
    dPre[iTimeInput][iFromInput] = dValue;
    return(dValue);}
```

```java
/** 7. vSetPost (Sets Postnumerando Cash Flows) */
public double          dSetPost(int iTimeInput, int iFromInput,
                                 int iToInput, double dValue){
  if (bGetData == true)
    { if (iMat2Idx[iFromInput][iToInput] == -1) return(0.);
        return(dPost[iTimeInput][iMat2Idx[iFromInput][iToInput]]);
    }
    bDKCalculated = false;
    if (iMat2Idx[iFromInput][iToInput] == -1)
        {
        iMat2Idx[iFromInput][iToInput] = iNrIndex;

        iFrom[iNrIndex] = iFromInput;
        iTo[iNrIndex] = iToInput;
        ++iNrIndex;
        }
    dPost[iTimeInput][iMat2Idx[iFromInput][iToInput]] = dValue;
    return(dValue);}

/** 8. vSetPij (Sets Probabilities ) */
public double          dSetPij(int iTimeInput, int iFromInput,
                                int iToInput, double dValue){
  if (bGetData == true)
    { if (iMat2Idx[iFromInput][iToInput] == -1) return(0.);
        return(dPij[iTimeInput][iMat2Idx[iFromInput][iToInput]]);
    }
    bDKCalculated = false;
    if (iMat2Idx[iFromInput][iToInput] == -1)
        {
        iMat2Idx[iFromInput][iToInput] = iNrIndex;
        iFrom[iNrIndex] = iFromInput;
        iTo[iNrIndex] = iToInput;
        ++iNrIndex;
        }
    dPij[iTimeInput][iMat2Idx[iFromInput][iToInput]] = dValue;
    return(dValue); }

/** 9. vSetDisc (Sets Discounts) */
public double          dSetDisc(int iTimeInput, int iFromInput,
                                 int iToInput, double dValue){
  if (bGetData == true)
    {
        return(dDisc[iTimeInput][iFromInput]);
    }
    bDKCalculated = false;
    dDisc[iTimeInput][iFromInput] = dValue;
    return(dValue);}
/* $$$$$$$$$$$$$$$$$$$$$$$$$$$$$ 10. MAIN MARKOV CALCULATOR $$$$$$$$$$$$$$$$$$$$$*/

/* The following routine calculates the mathematics reserve, which is called
   also premium reserve or "Deckungskapital" DK in German */

public double dGetDK(int iTimeInput, int iStateInput){

/*===========================================================================*/
    if (bDKCalculated == true)
        return (dV[iTimeInput][iStateInput]);
    bDKCalculated = true;

/*==== 0. Reset =============================================================*/
    for (iT = 0; iT < iMaxTimes; ++ iT)
    {
        for(iC1= 0; iC1 < iMaxStates; ++ iC1)
            dV[iT][iC1] = 0.;
    }
/*==== 1. Recursion =========================================================*/
```

```
for(iT = iStart -1; iT >= iStop; --iT)
{
  /* 1. fill dVTemp with Pre payments */
  for(iC1= 0; iC1 < iStates; ++iC1){dVTemp[iC1]= 0.;}
  /* 2. Post payments and the mathematical reserve              */
  for(iC1= 0; iC1 < iNrIndex; ++iC1)
    {dVTemp[iFrom[iC1]] +=  dPij[iT][iC1] * (dPost[iT][iC1]
                                          + dV[iT+1][iTo[iC1]]);}
  /* 2a. Pre payments and discount */
  for(iC1= 0; iC1 < iStates; ++iC1){dV[iT][iC1]= dVTemp[iC1]*dDisc[iT][iC1]
              }                                 + dPre[iT][iC1];}
  return (dV[iTimeInput][iStateInput]);
}}
```

References

[Bau91] H. Bauer, *Wahrscheinlichkeitstheorie* (De Gruyter, Berlin, 1991)

[Bau92] H. Bauer, *Mass- und Integrationstheorie* (De Gruyter, Berlin, 1992)

[BGH+86] N.L. Bowers, H.U. Gerber, J.C. Hickman, D.C. Jones, C.J. Nesbitt, *Actuarial Mathematics* (Society of Actuaries, Schaumburg, 1986)

[BP80] D. Bellhouse, H. Panjer, Stochastic modelling of interest rates with applications to life contingencies. J. Risk Insur. **47**, 91–110 (1980)

[BS73] F. Black, M. Scholes, The pricing of options and corporate liabilities. J. Polit. Econ. **81**, 637–654 (1973)

[Büh92] H. Bühlmann, Stochastic discounting. Insur. Math. Econ. **11**(2), 113–127 (1992)

[Büh95] H. Bühlmann, Life insurance with stochastic interest rates. in *Financial Risk in Insurance*, ed. by G. Ottaviani (Springer, Heidelberg, 1995), pp. 1–24

[CHB89] J. Cox, C. Huang, Option pricing theory and its applications. in *Theory of Valuation*, ed. by S. Bhattachary, G.M. Constantinides (Rowman and Littlefield Publishers, lanham, 1989), pp. 272–288

[CFO08] CFO Forum, Market Consistent Embedded Value, http://www.cfoforum.nl/embedded_value.html

[CIR85] J. Cox, J. Ingersoll, S. Ross, A theory of term structure of interest rates. Econometrica **53**, 385–408 (1985)

[CW90] K.L. Chung, R.J. Williams, *Introduction to stochastic Integration*, 2nd edn. (Birkhäuser, 1990)

[DAV09] DAV Unterarbeitsgruppen "Rechnungsgrundlagen der Pflegeversicherung" und "Todesfallrisiko". *Herleitung der Rechnungsgrundlagen DAV 2008 P für die Pflegerenten(zusatz)versicherung*, und *Raucher- und Nichtrauchersterbetafeln für Lebensversicherungen mit Todesfallcharakter*, und *Herleitung der Sterbetafel DAV 2008 T für Lebensversicherungen mit Todesfallcharakter. Blätter der DGVFM*, 30/1:31–140, 141–187, 189–224, 2009

[Doo53] J.L. Doob, *Stochastic Processes* (Wiley, New York, 1953)

[Dot90] M. Dothan, *Prices in Financial Markets* (Oxford University Press, New York, 1990)

[DS57] N. Dunford, J.T. Schwartz, *Linear Operators Part 1: General Theory* (Wiley-Interscience, New York, 1957)

[Duf88] D. Duffie, *Security Markets: Stochastic Models* (Academic Press, San Diego, 1988)

[Duf92] D. Duffie, *Dynamic Asset Pricing Theory* (Prinston University Press, Princeton, 1992)

[DVJ88] D.J. Daley, D. Vere-Jones, *An Introduction to the Theory of Point Processes* (Springer, Heidelberg, 1988)

[Fel50] W. Feller, *An Introduction to Probability Theory and its Applications* (Wiley, New York, 1950)

M. Koller, *Stochastic Models in Life Insurance*, EAA Series,
DOI: 10.1007/978-3-642-28439-7, © Springer-Verlag Berlin Heidelberg 2012

[Fis78] M. Fisz, *Wahrscheinlichkeitsrechnung und mathematische Statistik* (Deutscher Verlag der Wissenschaften, Berlin, 1978)

[Ger95] H.U. Gerber, *Life Insurance Mathematics*, 2nd edn. (Springer, Berlin, 1995)

[HK79] J.M. Harrison, D. Kreps, Martingales and multiperiod security markets. J. Econ. Theory **20**, 381–401 (1979)

[HN96] O. Hesselager, R. Norberg, On probability distributions of present values in life insurance. J. Insur. Math. Econ. **18**(1), 135–142 (1996)

[Hoe69] J.M. Hoem, Markov chain models in life insurance. Blätter der Deutschen Gesellschaft für Versicherungsmathematik **9**, 91–107 (1969)

[HP81] J.M. Harrison, S.R. Pliska, Martingales, stochastic integrals and continuous trading. Stoch. Process. Appl. **11**, 215–260 (1981)

[Hua91] C. Huang, Lecture notes on advanced financial econometrics. Technical Report, (Sloan School of Management, MIT, Massachusetts, 1991)

[Hul97] J.C. Hull, *Options, Futures and Other Derivatives* (Prentice Hall, 1997)

[IW81] N. Ikeda, S. Watanabe, *Stochastic Differential Equations and Diffusion Processes* (North-Holland, 1981)

[KP92] P. Kloeden, E. Platen, Numerical Solution of Stochastic Differential Equations. in *Applications of Mathematics*, vol 23 (Springer, Heidelberg, 1992)

[KS88] I. Karatzas, S.E. Shreve, *Brownian Motion and Stochastic Calculus* (Springer, New York, 1988)

[Mol92] C.M. Moller, Numerical evaluation of markov transition probabilities based on the discretized product integral. Scand. Actuar. J. **1**, 76–87 (1992)

[Mol95] C.M. Moller, A counting process approach to stochastic interest. Insur. Math. Econ. **17**, 181–192 (1995)

[NM96] R. Norberg, C.M. Moller, Thiele's differential equation by stochastic interest of diffusion type. Scand. Actuar. J. **1**, 37–49 (1996)

[Nor90] R. Norberg, Payment measures, interest and discounting. Scand. Actuar. J. 14–33 (1990)

[Nor91] R. Norberg, Reserves in life and pension insurance. Scand. Actuar. J. **1**, 3–24 (1991)

[Nor92] R. Norberg, Hattendorff's theorem and thiele's differential equation generalized. Scand. Actuarial J. **1**, 2–14 (1992)

[Nor94] R. Norberg, Differential equations for higher order moments of present values in life insurance. Insur. Math. Econ. 171–180 (1994)

[Nor95a] R. Norberg, Stochastic calculus in actuarial science. Working paper, Laboratory of Actuarial Mathematics University of Copenhagen, 1995

[Nor95b] R. Norberg, A time-continuous markov chain interest model with applications to insurance. J. Appl. Stoch. Models Data Anal. **11**, 245–256 (1995)

[Nor96a] R. Norberg, Addendum to hattendorff's theorem and thiele's differential equation generalized. Scand. Actuar. J. 2–14 (1996)

[Nor96b] R. Norberg, Bonus in life insurance: principles and prognoses in a stochastic environment. Working paper, Laboratory of Actuarial Mathematics University of Copenhagen, 1996

[Nor98] R. Norberg, Vasicek beyond the normal. Working paper, Laboratory of Actuarial Mathematics University of Copenhagen, 1998

[Par94a] G. Parker, Limiting distributions of the present value of a portfolio. ASTIN Bull. **24**(1), 47–60 (1994)

[Par94b] G. Parker, Stochastic analysis of a portfolio of endowment insurance policies. Scand. Actuar. J. **2**, 119–130 (1994)

[Par94c] G. Parker, Two stochastic approaches for discounting actuarial functions. ASTIN Bull. **24**(2), 167–181 (1994)

[Per94] S.A. Persson, Pricing Life Insurance Contracts under Financial Uncertainty, Ph.D. Thesis, Norwegian School of Economics and Business Administration, Bergen, 1994

[Pli97] S.R. Pliska, *Introduction to Mathematical Finance. Discrete Time Models* (Blackwell Publishers, Malden, 1997)

[Pro90] P. Protter, Stochastic Integration and Differential Equations. in *Applications of Mathematics*, vol 21 (Springer, Heidelberg, 1990)

[PT93] H. Peter and J.R. Trippel, Auswertungen und Vergleich der Sterblichkeit bei den Einzelkapitalversicherungen der Schweizerischen Lebensversicherungs- und Rentenanstalt in den Jahren 1981–1990. Mitteilungen der Schweiz. Vereinigung der Versicherungsmathematiker pages 23–44, 1993

[RH90] H. Ramlau-Hansen, Thiele's differential equation as a tool in product developpement in life insurance. Scand. Actuar. J. 97–104 (1990)

[Rog97] L.C.G. Rogers, The potential approach to the term structure of interest rates and foreign exchange rates. Math. Finance 157–176 (1997)

[Vas77] O. Vasicek, An equilibrium characterisation of the term structure. J. Financial Econ. **5**, 177–188 (1977)

[WH86] H. Wolthuis, I. Van Hoek, Stochastic models for life contingencies. Insur. Math. Econ. **5**, 217–254 (1986)

[Wil86] A.D. Wilkie, Some applications of stochastic interest models. J. Inst. Actuar. Stud. Soc. **29**, 25–52 (1986)

[Wil95] A.D. Wilkie, More on a stocastic asset model for actuarial use. Br. Actuar. J. **1**, 777–964 (1995)

[Wol85] H. Wolthuis, Hattendorf's theorem for a continuous-time markov model. Scand. Actuar. J. **70**, (1985)

[Wol88] H. Wolthuis, Savings and Risk Processes in Life Contingencies, Ph.D. Thesis, University of Amsterdam, 1988

Notation

Index

M. Koller, *Stochastic Models in Life Insurance*, EAA Series,
DOI: 10.1007/978-3-642-28439-7, © Springer-Verlag Berlin Heidelberg 2012